Warship 2023 Conference

Delivering Submarine Capability and Availability with Agility and Pace

Bath, United Kingdom
31 May - 1 June 2023

ISBN: 978-1-7138-9565-7

Printed from e-media with permission by:

Curran Associates, Inc.
57 Morehouse Lane
Red Hook, NY 12571

CURRAN ASSOCIATES INC. proceedings.com

Some format issues inherent in the e-media version may also appear in this print version.

The Royal Institution of Naval Architects
8-9 Northumberland Street
London, WC2N 5DA
United Kingdom

Phone: 020 7235 4622
Fax: 020 7259 5912

publications@rina.org.uk

Additional copies of this publication are available from:

Curran Associates, Inc.
57 Morehouse Lane
Red Hook, NY 12571 USA
Phone: 845-758-0400
Fax: 845-758-2633
Email: curran@proceedings.com
Web: www.proceedings.com

160 YEARS OF HISTORY,
KNOWLEDGE AND EXPERIENCE

SALUM ET CARINAE PIGNORA VITAE

THE ROYAL INSTITUTION
OF NAVAL ARCHITECTS

WARSHIP 2023 CONFERENCE

Delivering Submarine Capability and Availability with Agility and Pace

31st May - 1st June 2023 Guildhall, Bath

Sponsored by:

BMT Ansys

THE ROYAL INSTITUTION
OF NAVAL ARCHITECTS

Warship Conference

31st May – 1st June 2023
Guildhall, Bath

THE ROYAL INSTITUTION OF NAVAL ARCHITECTS
8-9 Northumberland Street
London WC2N 5DA

Telephone: 020 7235 4622
Fax: 020 7259 5912

ISSN No: 2752-6569

CONTENTS

IN-PROCESS ULTRASONIC INSPECTION DURING FUSION WELDING OF LARGE-SCALE FABRICATIONS 1
C N MacLeod, D Lines, R Vithanage, M Vasilev, C Loukas, N Sweeney, E Foster, E Mohseni, Y Javadi, G Dobie, S G Pierce, A Gahagan, P Applequist, University of Strathclyde, UK, N King, R Whitmore, Babcock International Group, UK, A Burnett, C Murray, P Robinson, Cavendish Nuclear, UK, B Holt, M Symington, J Allan, G Little, Altrad Babcock, UK, M Smart, D L Ayres, J Leatherland, University of Sheffield, UK

USING THE NAVAL SUBMARINE CODE (NSUBC) AS MEANS FOR SUBMARINE ASSURANCE DURING DESIGN, CONSTRUCTION AND OPERATION 7
Ch von Oldershausen, DNV, Germany

SUBMARINE VULNERABILITY REDUCTION: WYVERN® AND PURPLE FIRE 15
C S Ritchie and J S Schofield, Survivability Consulting Limited, UK A E Walchester, and D Patten, BMT, UK

CAPABILITY VERSUS AVAILABILITY IN SUBMARINE DESIGN? 27
D Andrews, University College London

WAKES OF SURFACE-PIERCING CYLINDERS 43
D Potts and Z Leong, Australian Maritime College, Australia J Binns and A Skvortsov, Defence Science and Technology Group, Australia H Marcollo, AMOG Consulting, Australia

CAPABILITY, AVAILABILITY, COST; THE TRADE AND THE BALANCE 55
G Wainwright, Babcock International, UK and R Chaplin, BAE Systems Submarines, UK

MBSE DRIVING DOWN THROUGH LIFE COST IN NAVAL APPLICATIONS 69
I Tolchinsky, and A Davenport, Ansys, USA

INCLUSIVE SUBMARINE DESIGN CONSIDERATIONS 77
J A Dutton, BMT Canada, Canada

THE EFFECT OF ORTHOTROPIC STEEL ON PRESSURE HULL COLLAPSE 85
J J Reijmers and A Zambon, Nevesbu b.v., The Netherlands

LEVERAGING CROSS-INDUSTRY APPROACHES TO ENABLE SUBMARINE DISPOSAL THROUGH IMPROVED DESIGN 93
K Dowrick and J Sheasby, BMT Ltd

EXPERIMENTAL DEVELOPMENTS IN ASSESSING THE SEAKEEPING BEHAVIOUR OF AN UNDERWATER VEHICLE OPERATING ON THE SURFACE 103
M Courdier, Z Leong, J Duffy, Australian Maritime College, Australia
J Binns, A Conway, Defence Science and Technology Group, Australia

APPLICATION OF SYSTEMS ENGINEERING (SE) WITHIN THE NAVAL ARCHITECTURE DOMAIN, WITH SPECIFIC APPLICATION OF MODEL BASED SYSTEMS ENGINEERING (MBSE) 117
M E Flexer and B Bosley, BAE Systems – Submarines, UK

DELIVERING SUBMARINE AVAILABILITY THROUGH DESIGN 131
R D Millard and A E Walchester, BMT, UK

EXPLORATIONS OF NOVEL NUCLEAR PROPULSION FOR SUBMARINES 141
R J Pawling, UCL Department of Mechanical Engineering, UK

PATROLLING THE METAVERSE 159
R Perez Fernández, and M Toman, SIEMENS DI SW, Spain

PRESSURE HULL CORROSION MODELLING AND ASSESSMENT 171
S Dawe, Babcock SSE, UK

DERIVING SUBMARINE SIGNATURE REQUIREMENTS 181
W J Wood, BMT Canada, Canada

IN-PROCESS ULTRASONIC INSPECTION DURING FUSION WELDING OF LARGE-SCALE FABRICATIONS

C N MacLeod, D Lines, R Vithanage, M Vasilev, C Loukas, N Sweeney, E Foster, E Mohseni, Y Javadi, G Dobie, S G Pierce, A Gahagan, and **P Applequist,** Centre of Ultrasonic Engineering (CUE), Department of Electronic & Electrical Engineering, University of Strathclyde, UK, **N King**, and **R Whitmore**, Babcock International Group, UK, **A Burnett**, **C Murray**, and **P Robinson**, Cavendish Nuclear, UK, **B Holt, M Symington, J Allan,** and **G Little**, Altrad Babcock, UK, **M Smart, D L Ayres,** and **J Leatherland**, Nuclear Advanced Manufacturing Research Centre (NAMRC), University of Sheffield, UK

SUMMARY

Traditionally, welding and inspection of welds on large-scale fabrications, such as submarine pressure hull components, are distinctly separate sequential manufacturing processes, which limits productivity and throughput along with increasing re-work if defects are detected post-build. The Advanced Welding Equipment System Inspection and Monitoring (AWESIM) programme introduces ultrasonic phased array inspection directly during welding to allow detection of imperfections and flaws as they occur, thus reducing rework, repair and delivering high-quality welds right, first time.

Results of this work on sub-scale plate and pipe demonstrators are presented with both intentionally controlled and unintentional defects. The signal to noise ratio and sensitivity of the imaging and inspection system is measured, evaluated, and quantified offering benefits such as reduced rework, schedule slippage and cost.

Babcock International Group will highlight the benefit it aspires to achieve from deploying this technology in its work on submarine component fabrication and submarine maintenance for the UK MoD.

1. INTRODUCTION

Babcock International Group has a long-standing history of providing submarine components and systems to the Royal Navy and in maintaining submarines at the Devonport Royal Dockyard and HMNB Clyde.

The production of high integrity welded joints is a critical part of submarine manufacturing. Quality control measures including workmanship inspection, surface and volumetric non-destructive testing are used to ensure weld acceptance standards are achieved.

Welding is a complex process requiring the control of many variables to enable defect free joints to be produced. Many welding processes in submarine component fabrication are performed manually; however, mechanised, or automated welding processes are used, where possible, to minimise the variations associated with manually deployed welding activities. Nevertheless, even low defect rates on large thick section multi pass automated welds can lead to significant repair times and consequential adverse programme/schedule impact.

Babcock International Group, together with the University of Strathclyde, Centre for Ultrasonic Engineering and other industrial partners, have developed an in process ultrasonic inspection technique capable of detecting the presence of defects in real time during the deposition of multi-pass welds. Benefits of early detection include significantly reduced repair times, minimisation of schedule impact and the ability to carry out immediate root cause investigation.

2. DEVELOPMENT TO DATE

Typically, volumetric weld inspection is carried out after completion of welding and the joint has cooled to ambient temperature [1,2], a further delay is often required to allow for hydrogen diffusion. In multi-pass welds this approach can lead to weld defects in early passes not being detected until final inspection. On large scale components this can result in a delay of several weeks before a quality issue is known and makes the rectification process more extensive and complex.

A programme of work was carried out in 2017 to 2019 at the Advanced Nuclear Research Centre (ANRC), University of Strathclyde, Centre for Ultrasonic Engineering to develop a Phased Array Ultrasonic Testing (PAUT) technique capable of detecting the presence of welding defects during the weld deposition process

[3]. In 2020, further work was undertaken using UK Government match funding (from the Department of Business Energy and Industrial Strategy, in Phase 2b of the Nuclear Innovation Programme Advanced Materials and Manufacturing fund) [4]. This work refined the system to allow for high-temperature dry-coupled PAUT inspection in conjunction with Gas Tungsten Inert Welding (GTAW) deposition [5]. This work also included an investigation of continuous weld process monitoring using a laser profile sensor, a high-definition camera, a weld arc sound sensor, and weld process control monitoring in parallel to detect weld defects in real time, performed by the Nuclear Advanced Manufacturing Research Centre at the University of Sheffield.

The development programme has addressed a range of technical challenges including:

- Requirement for an automated system
- Temperature gradient compensation
- Electromagnetic interference suppression
- High temperature, dry coupling methodology to avoid the need for wet/gel acoustic couplant
- Masking of reflections from unfused weld prep faces
- Compensation for variation in the thickness of the joined materials

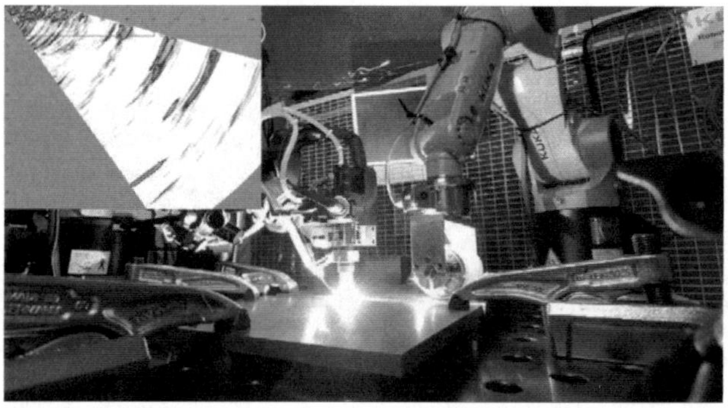

Figure 1: Synchronous flat plate GTAW and inspection trials using 6-axis robot arms [3]

Following hardware and software development activities, a comprehensive programme of system proving on test plates was carried out to confirm the ability of the probe to detect typical 'real' defect types including lack of side wall and inter run fusion, porosity and solid inclusions. A series of test plates were produced with adjustments to welding parameters, torch position and levels of surface contamination to produce the necessary conditions required to create typical production defects. Defect types were confirmed with macro sections.

For an industrially relevant demonstration, a tubular test piece, (406mm OD, 18mm wall thickness) with a single vee weld preparation, was utilised to enable a range of artificial and 'real' defect types to be randomly distributed throughout the joint. All defects were successfully detected in real time during the weld deposition process. Signal amplitude, and signal to noise ratio and spatial resolution was assessed and compared with conventional commercially available hardware with positive results.

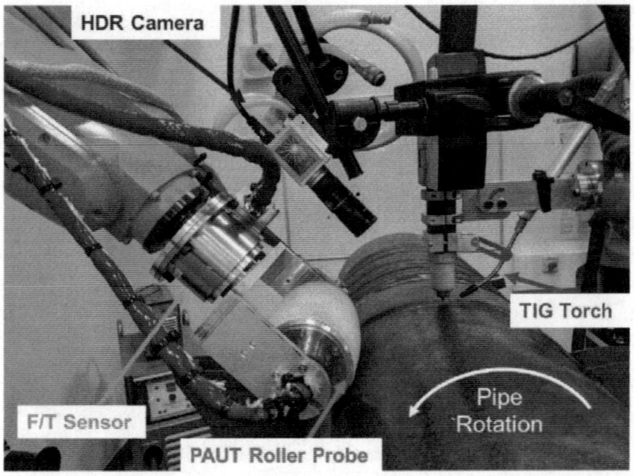

Figure 2: PAUT roller probe inspection of tubular test piece [3]

3. CURRENT PROGRAMME

A further development programme has just commenced. The aim of the following two projects is to take the technology arising from AWESIM into commercial use.

3.1 PROJECT 1: SYNCHRONOUS WELDING AND PAUT INSPECTION OF SEAM WELDS

Synchronous welding and PAUT roller probe inspection of longitudinal and circumferential seam welds in large scale components is particularly relevant to submarine construction and component manufacture. Material types used in this project will include C Mn structural and Low Alloy steels.

The plan is to investigate practical deployment requirements associated with large scale typical production assemblies. The work will initially focus on relatively simple longitudinal and circumferential seam welds before moving on to more complex geometries including through wall penetrations. The intent is to investigate full automated synchronous operation of a robotic controlled GMAW welding and PAUT roller probe inspection process.

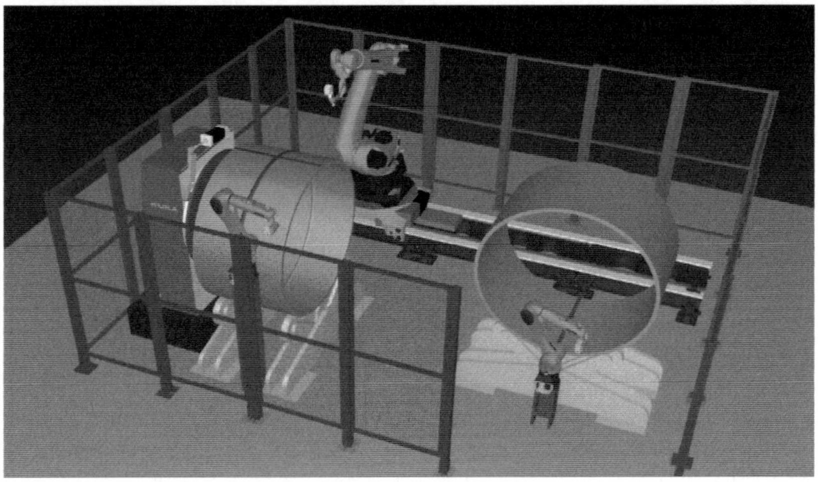

Figure 3: GMAW robotic welding cell for longitudinal / circumferential PAUT roller probe trials

3.2 PROJECT 2: ORBITAL PAUT ROLLER PROBE INSPECTION OF FIXED PIPEWORK INSTALLATIONS

Development of PAUT roller probe deployment methods using orbital track / bracelet deployment will be undertaken to enable the technology to be employed on inspecting in-situ pipe welds. It is intended that this technique will be used in factory fabrication of pipe modules as well as during submarine maintenance involving pipes with an external diameter >150mm. Material types used in this project will include C Mn structural, Low Alloy and Austenitic Stainless steels.

4. BENEFITS OF IN PROCESS PAUT INSPECTION

In process PAUT will provide a significant range of benefits to high integrity multi pass welding activities on submarines and their components.

At contract start-ups, defect rates tend to be higher when operators are learning new component set-ups, coping with new equipment and weld procedures as well as trying to maintain schedule requirements. The ability to closely monitor the quality of deposited weld metal at the early stages of a new project will be particularly advantageous.

A summary of other key benefits include:

- Minimisation of repair activities leading to improved schedule adherence, reduction in energy use and minimisation of material wastage
- The ability to carry out immediate root cause investigations – often made difficult due to time delay from completing welding to carrying out final volumetric NDT
- Early identification of equipment malfunctions
- Easier introduction of new operators due to in process monitoring
- Technology can be used to support the weld procedure development process
- Provision of enhanced digital records of weld quality history
- Can be incorporated into a formal quality control surveillance programme
- Minimisation of possible thermal damage and distortion / residual stress
- Avoidance of the need for gel / liquid acoustic couplant use and clean up

It is important to note that, presently, the technology is not intended to be used as a final certification of weld compliance to code. The intent is to use the technology to assess weld quality during the joining process, and, therefore, to provide complementary evidence in support of final code compliant weld inspection. In future we envisage further work will be performed with organisations such as ASME, ISO etc. with the aim of ensuring that future code updates allow for the use of in process thermally compensated PAUT for final weld certification.

7. CONCLUSIONS

This paper and the associated presentation provide an insight into the development of an in-process PAUT inspection system which has the potential to transform welding and associated NDT practices. An explanation of what has been achieved to date and current development plans are highlighted with a focus on projects that are aimed at improving productivity in the production and / or maintenance of high integrity components for submarines.

8. ACKNOWLEDGEMENTS

The authors wish to acknowledge support from the UK Government, our respective organisations and our respective teams who have committed their time and resources so enthusiastically to make this programme of work a reality.

9. DISCLAIMER

The opinions expressed are solely those of the authors.

10. REFERENCES

1. ISO19285:2017, 'Non-destructive testing of welds - Phased array ultrasonic testing (PAUT) - Acceptance levels', *International Organization for Standardization*, 2017

2. ISO17640:2017, 'Non-destructive testing of welds - Ultrasonic testing - Techniques, testing levels, and assessment', *International Organization for Standardization*, 2017.

3. VITHANAGE, R K W., et al., 'A phased array ultrasound roller probe for automated in-process / inter-pass inspection of multi-pass welds', *IEEE Transactions on Industrial Electronics, Volume: 68 Issue: 12*, 08 December 2020.

4. The Engineer, 14 July 2020, https://www.theengineer.co.uk/content/news/babcock-s-awesim-project-could-help-reduce-cost-of-nuclear-new-build.

5. BURNETT, A J., 'Real-time weld defect identification', *Nuclear Engineering International*, April 2022.

9. AUTHORS' BIOGRAPHY

Dr Anthony Burnett holds the current position of Senior Engineering Manager at Cavendish Nuclear Ltd (a wholly owned subsidiary of Babcock International Group). He is responsible for co-ordinating and leading the development of several innovation projects including the technology arising from the AWESIM project. Previously, he was the Head of Innovation and Technology for Cavendish Nuclear and is still active in this area. He is a mechanical engineer by profession with a technical background that has mainly focused on the field of nuclear decommissioning.

Prof. Charles MacLeod is a Professor in the Centre for Ultrasonic Engineering at the University of Strathclyde. He is the Babcock International Group / Royal Academy of Engineering Chair in Sensor-Driven Automated Welding. He is director of the Sensor Enabled Automation & Robotics Control Hub (SEARCH), a £24M research innovation and technology transfer laboratory. Aligned to this, he is Course Director of the Autonomous Robotic Intelligent Systems (ARIS) MSc. His research is primarily associated with the inspection and manufacturing enhancement of high-value assets and components, where he leads several high-profile inter-disciplinary in-process welding and metal additive research projects.

USING THE NAVAL SUBMARINE CODE (NSUBC) AS MEANS FOR SUBMARINE ASSURANCE DURING DESIGN, CONSTRUCTION AND OPERATION

Ch von Oldershausen, DNV, Germany

SUMMARY

Submarine Safety is a complex, multi-dimensional task covering all technical areas of the submarine platform. The importance of managing safety is therefore imperative at every major phase throughout the lifecycle of the submarine. Ensuring that this objective is achieved requires a robust and well-proven assurance process. Consequently, the Naval Submarine Code (NSubC) often in conjunction with classification society rules, is increasingly being used by navies as a powerful process and assurance tool, bearing in mind however, this is a process depending on the requirements of the respective Naval Flag Administrations. The NSubC is a goal-based standard, which sets a minimum level of safety for submarines and provides a framework for further work and development in this field. It considers the ultimate safety objectives and allows for the consideration of alternative designs and technical solutions that meet the objective. The NSubC has been recently published by NATO as new ANEP-102.

1. SETTING THE SCENE

Haakonsvern Naval Base, Norway. A group of submariners from Norway and Germany leave the shelter of their quarters, braving the wind and rain typical of this part of Norway. They are on their way to the quay to muster before starting preparations on their U212CD class submarine, which is due to sail shortly on a mission into the North Atlantic. Today, this may still sound like a fictitious scenario, but with the signing of the cooperation agreement between the German and Norwegian navies to build a series of common submarines, which shall have their home base in Norway, this will become reality in the future.

The two NATO partners have joined forces and have ordered the construction of six new submarines from ThyssenKrupp Marine Systems in 2019. The submarines are a further development of the successful German U212A class. After delivery, four of the submarines will enter service with the Norwegian navy and two with the German navy. As in the cooperation between the Royal Netherlands Navy and the Belgian Navy, jointly maintaining and operating the frigates of the M-Class and in future common MCMVs, this bilateral collaboration will provide a number of synergies for both the Norwegian and German navy. The uniform design will allow Germany and Norway, for example, to conduct joint training units, exercises, missions and most of all joint maintenance and logistic facilities.

On the insistence of Norway, the so-called Naval Submarine Code (NSubC) will be used as a process of assurance for this international cooperation project. This code, which was developed specifically for submarines, was developed by the International Naval Safety Association (INSA) - an association of navies and classification societies that has the objective of establishing uniform verification codes for the design and construction of warships and submarines. These codes provide a goal-based safety assurance framework for naval ships. Using these codes navies have a sound mechanism to demonstrate that they have implemented an effective safety management system that, whilst based on international conventions such as SOLAS, has recognised the very different operating philosophy that naval vessels adopt dealing with the portfolio of risks that they are exposed to. The codes, i.e., the Naval Ship Code (NSC) and NSubC are published through NATO as ANEP 77 and ANEP 102 respectively.

2. SAFETY ASSURANCE

In terms of naval applications, shipbuilders and navies find themselves faced with many challenges throughout the whole project lifecycle from the conceptual phase of a navy vessel or submarine until the end of an often-extended operational lifetime. Apart from typical constraints, which include budget limitations, a critical dilution of in-house expertise for shipbuilding and design, a shrinking pool of qualified personnel and finally, increasing demands from the public for safety and environmental protection the project is often faced with specific requirements. Sufficient range of capabilities and reliable function of the naval vessel and its systems, an expanded and flexible operational profile, exploitation of latest technology, high safety levels for the crew on board and compliance with future environmental standards, must be met. Ensuring that these objectives are achieved requires a robust and well-proven assurance process. Consequently, safety assurance based on proven Classification processes and rules is increasingly being used by navies as a powerful tool, bearing in mind however, that unlike in commercial shipping this is an optional process depending on the requirements of the respective Naval Administrations. In some cases, additional technical standards need to be considered to achieve a broader level of assurance. This might be enhanced through the application of a barrier management risk-based safety case assessment approach to the platform and/or its systems.

Some recent submarine accidents like the disappearance of the ARA San Juan in the South Atlantic on 15 November 2017 or the sinking of the KRI Nangala, which went missing during a routine exercise in the Bali Sea on 21 April 2021, in both cases with a substantial number of lives lost, highlight the importance of having a high safety culture applied to submarines not only during operation but already during design and construction. The importance of managing safety is therefore imperative at every major phase throughout the lifecycle of the submarine - from design to construction, through to testing, commissioning for service, and the operation of the submarine. The objective must be to always assure a safety critical environment on board of a submarine.

A key player in this respect is the navy's Naval Flag Administration administering the overall Safety Management Process. The role to fulfil is similar to that of a national Flag State:

- Ensuring compliance of vessels under their regulatory remit for compliance with Naval Safety Policy
- Maintaining understanding of relevant technical standards against which naval vessels may be assessed, including:
    - Legislation (e.g. UN, IMO)
    - Other NATO documents (ANEPs, STANAGs)
    - International Standards (e.g. ISO, EN)
    - Classification Societies rules and regulations
    - IACS Unified Requirements
    - National legislative requirements (e.g. EC, Baltic Ice)
    - National Navy standards (e.g. MilSpec, Def Stan)
- Advising on changes to international criteria that may impact a navy

The tasks of the Naval Flag Administration can be delegated to a Recognised Oganisation to act on behalf of the Naval Flag Administration, e.g., Classification Societies. (NSubC: PT.3 CH.0 ANNEX C). The classification societies are considered qualified to act as a Recognized Organisation, for other organisations there is a qualification scheme (procedures to appoint and audit) available. Through the use of an independent process of verification outside the responsibility of the asset owner (Navy), the Naval Flag Authorities are able to show a robust and transparent safety management process.

3. DNV'S NAVAL ASSURANCE APPROACH

DNV's assurance processes for naval surface vessels and submarines provide two options. In addition to the conventional Naval Classification scheme, which is based on the application of DNV's Naval Rules, there is the alternative to apply DNV's Naval Technical Assurance approach. As the name of this service indicates, it comprises of a more generic approach to the complete assurance process with respect to Material Safety, not limited to the established understanding of Classification. The scope can vary but the process is similar to conventional Classification.

Apart from addressing technical needs DNV Naval Technical Assurance has also been developed to support a wide range of regulatory approaches, which amongst others include the International Naval Safety Association codes (ANEP-77 Naval Ship Code, ANEP-102 Naval Submarine Code) as the main control of safety performance requirements.

4. THE NAVAL SUBMARINE CODE (NSUBC)

The complex nature of designing, building, and testing a submarine requires a robust platform and safety framework to:

a) identify any safety issues during the concept phase
b) ensure that safety can be ascertained at the design phase
c) assure that the manufacturing work carried out meets the high-quality standard required
d) ensure the safety requirements and emergency systems are vigorously tested to provide maximum assurance.

All of the above can be achieved by applying the process of the safety assurance framework NSubC.

Like the NSC the NsubC is a goal-based standard or rather a process, which as outcome when implemented, sets a minimum level of safety for submarines, and provides a framework for further work and development in this field. The benefits of this goal-based standard are that it considers the ultimate safety objectives and allows for the consideration of alternative designs and technical solutions that meet these objectives. The Naval Submarine Code is currently gaining acceptance in a similar way as the Naval Ship Code (NSC), which has now reached a level of maturity through the support of several countries who have implemented the NSC on their existing naval fleets and new vessels.

The Spanish Navy is considering ANEP-102 as means of safety assurance for the through life support of their S-80 submarines. The Royal Netherlands Navy is similarly looking to apply the NsubC as assurance process for the replacement of their Walrus Class submarines. For the Norwegian/German U212CD project, the NsubC will be used as a process of assurance, as requested by Norway.

Structured in the same way as the other two naval codes the NSubC has three parts (Figure 1 below). Part 1 comprises the Regulations, with the content of the chapters stipulating submarine specific safety provisions and requirements. Part 2 is designated for Solutions but is not populated but instead is relying on the user selecting appropriate standards and Part 3 contains the Code Justification and Guidance. It is only Part 1 which is published as ANEP-102 and classified as NATO RESTRICTED. Part 3 is made available to INSA members only, who contribute to its development. It is foreseen that those nations may publish the Naval Submarine Code under their own cover for use on their specific projects and thus in doing so, they may choose to populate Part 2 with their own solutions.

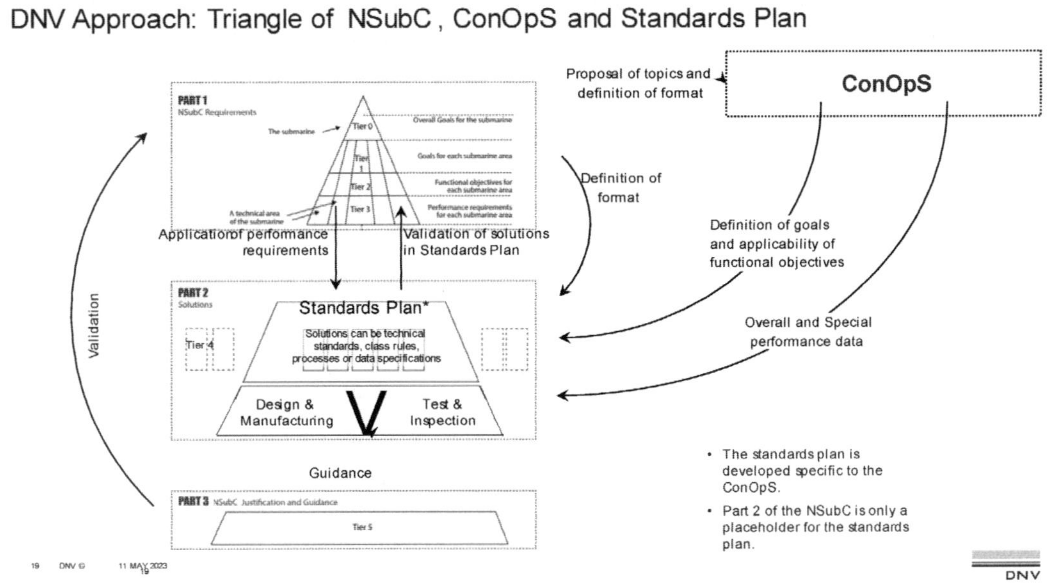

Figure 1: NSubC / Dependency of ConOpS, Standards Plan and Solutions

4.1 DEPENDENCY OF CONOPS, STANDARDS PLAN AND SOLUTIONS

The NSubC is not a "stand-alone" document as it requires project specific documents as prerequisites accompanying and/or initiating the safety assurance process.

4.1 (a) Concept of Operations Statement (ConOpS)

At the top level the Concept of Operations Statement (ConOpS) defines the submarine's function, operational characteristics, and areas. Typical aspects such submarine attributes, which include design life, dimensions, diving depths, speeds, range, endurance, operational area, payloads, Replenishment at Sea (RAS), anchoring and mooring, role specific operations (diver lock in/out), embarked persons and accommodation requirements but also survivability definitions, stipulating reserve of buoyancy, shock & hostile weapon impact, maneuvering envelope (speed and depth), collision, fire & flooding and last but not least the definition of the environment with requirements concerning meteorology and climatology (above surface), sea surface (interface), bathymetry and oceanography (below surface), geotechnical (bottom and bank conditions), human caused environment (berthing, bottoming, towing & salvage, signature, launching, noise & vibration) and many more aspects should be defined in the ConOpS.

The definition of the ConOpS should also include an appraisal process of the limitations so as to avoid any issues surfacing during the design and build phase of the submarine. Full awareness of any limitations or restrictions associated with the submarines use and modes of operation is a crucial criterion before embarking on the design and engineering phase.

Last but not least the ConOpS should also stipulate the envisaged survey and maintenance philosophy with the associated schedules. This includes completeness and efficiency of the inspection and survey regime during the operational lifetime of the submarine.

As such the ConOpS is the formal basis for the assurance process against the NSubC. The document may, however, be a sub-set of a more detailed military CONOPS and for practical reasons shall be split in two parts, i.e., a classified and a non-classified part.

4.1 (b) Standards Plan

Following the ConOpS and to be regarded as the foundation of the NSubC implementation is the Standards Plan. This document is usually developed or issued by the Naval Flag Administration and is specifically tied to the ConOpS. The Standards Plan is specific to the ConOpS and changes when the ConOpS is modified. It provides a Solution to every safety Performance Requirement of the NSubC, which can range from "vessel level" down to "component level".

4.1 (c) Solutions

Each Solution addressing a Performance Requirement is defined by a set of specific Standards. These Solutions are no longer goal based as they specifically define a technical solution and the measure of verification. The NSubC stipulates that the submarine, its systems, and equipment are to comply with, and be approved in accordance with justified classification society's rules or other suitable justified standards to facilitate verification of the Performance Requirements. The submarine builder should to a great extent be involved in the stipulation of the technical solution and provide information regarding the impact on the design and construction of the submarine.

DNV's Submarine Classification Rules, as part of DNV 's RU-NAV Rule Set, are structured in such a way that they provide Solutions to the requirements of the NSubC as defined in Part 1.

Once the Standards Plan with its respective Solutions is established a thorough process of Verification and Validation is needed to ensure each Performance Requirement is met. By translating the Performance Requirements of the NSubC into a set of Verification Plans every Solution to a Performance Requirement is split up to address three levels, i.e., submarine level, system level and component level. Any deviations from the NSubC or noncompliance will require an Acceptance by the Naval Flag Authority and must be justified and documented in the Standards Plan.

5. VERIFICATION AND VALIDATION

Following the Systems Engineering V-Modell the Assurance Process applied by DNV trails the V down during the various steps of the design phase and the upwards again during the integration through a verification phase with the objective of understanding the state of the platform design and construction, determining whether it is meeting its requirements (Figure 2).

In accordance with ISO/IEC 15288 (Systems and software engineering — System life cycle Processes) Validation is an activity of checking that requirements defined for development, define the system that the customer really wants. Verification is a test of a system to prove that it meets all its specified requirements at a particular stage of its development.

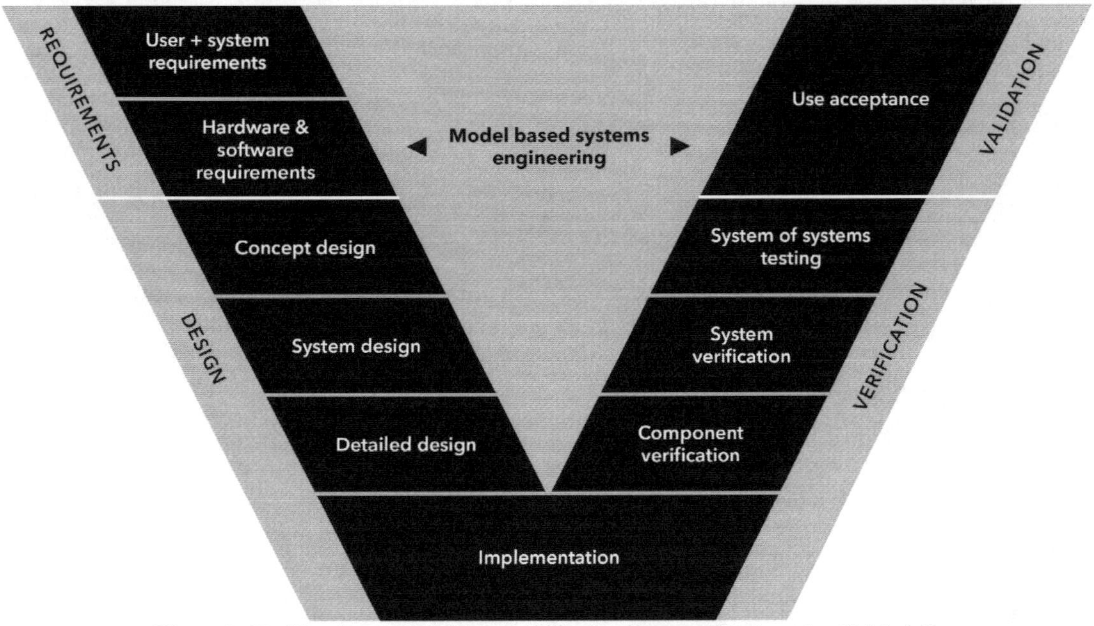

Figure 2: Verification and Validation following Systems Engineering V-Modell

5.1 CONCEPT PHASE

During the concept phase the functional/capability requirements of a submarine are determined, taking account of technological, commercial and market parameters. The concept phase will typically include the selection of the regulatory regimes, class rules, standards, codes, and certifications, e.g., the NSubC, to be used in the design and construction of the submarine, resulting in a binding schedule of certificates (Standards Plan) listing all performance and quality records to be provided. DNV usually supports the project during this phase by advising on submarine safety considerations through the provision of a robust safety assurance process in line with the functional/capability requirements of the navy.

5.2 SYSTEMS AND DETAILED DESIGN PHASE

The subsequent design phase requires that a determined number of plans and documents are submitted for technical appraisal against the agreed standards and class rules (Standards Plan). Typical aspects to be considered when assessing submarine safety include the structural integrity of the platform, machinery, electrical equipment, and safety systems. The objective of these design reviews and the plan approval is to ensure that the platform and systems design meet the contractually agreed standards for safety and quality. The design assessment covers many aspects including: ensuring that the material selected is suitable, a review of risk assessments (e.g. FMEA), and an appraisal of emergency procedures.

5.3 CONSTRUCTION PHASE, MATERIAL AND COMPONENT VERIFICATION

The involvement of classification includes not only the survey of construction at the shipyard but also the inspection of material, equipment, and systems at the suppliers' premises. It might also involve a welding certification by DNV prior to the commencement of manufacturing with a mandatory workshop approval at the shipyard and of sub-contractors. This includes certification of welders, approval of materials and welding consumables, and the implementation of a Welding Procedure Specification (WPS) and a Welding Procedure Qualification Record (WPQR).

The assurance process ensures the material and components used meet the technical specification and classification requirements, and furthermore a confirmation that the construction is in accordance with the design.

5.4 SYSTEMS VERIFICATION AND TESTING PHASE

During the testing phase, which includes Factory Acceptance Tests (FATs), Shore Base Integration Tests (SBITs), Platform Integration Tests (PITs), Harbour Acceptance Test (HAT) and Sea Trials with diving tests, DNV witnesses the functioning of all safety relevant systems, assuring that the mitigating measures for safety risks identified during the design and construction phase have been implemented. DNV submarine experts will also participate in the deep diving tests of the submarine.

5.5 OPERATIONAL PHASE

Through the submarine's life, periodic inspections are conducted by DNV's experienced submarine surveyors, assuring that compliance with the standards agreed during the previous phases (NSubC and its safety goals) are maintained. The inspections vary according to the position of the submarine within the classification cycle, which typically is five years. The length of the classification cycle can also be adjusted to match overhaul and/or maintenance periods, for example the three South African HDW-Class 209/1400mod submarines follow a three-year classification cycle.

There are different types of periodical inspections:

- Annual Surveys
- Intermediate survey after the 2nd year
- Renewal survey after completion of the "classification cycle", which includes a dry docking to inspect the hull and appendices
- Special survey, for example due to a damage repair, and which requires the engagement of the classification society to assess the situation and provide advice on the repair to maintain classification requirements.

The results of the inspections are documented in survey reports, which detail any defects and/or conditions of class which must be rectified within a certain period.

6. CONCLUSION

The NSubC contributes to the Naval Flag Administrations overall Safety Management System. The NSubC is a tool for the Naval Flag Administration to demonstrate that the submarine is materially safe to operate and to establish procedures for maintaining this condition. Competence of crew, workplace health & safety, environmental protection etc. are separate items but are supported by operational procedures, training, and documentation requirements.

Applying the goal-based safety assurance framework NSubC and using the naval classification services of DNV as a robust and well-proven assurance process provides:

- an efficient way of ensuring a safe, reliable and ready for duty naval platform in compliance with all relevant standards and regulation
- transparency evidence of an adequate process for control and the achieved level of safety
- the application of continuously maintained and updated assurance rule set

- a safe and reliable design in compliance with all relevant national & international standards

- assurance of high quality of materials and components and a high fabrication quality incl. welding procedure etc.

- enhances the ability of naval authorities and/or procurement organisations to focus on the military aspects of the naval platform

- the benefit of an organisation used to make decisions on acceptance limits and the proper interpretation of formal requirements as an independent and impartial Recognised Organisation

- a cost-effective approach to safety and quality through clearly agreed standards, procedures, requirements, and specifications

and thus demonstrates the implementation of an effective safety management system of a submarine.

7. AUTHOR'S BIOGRAPHY

Christian von Oldershausen started his professional career in the German Navy sailing on destroyers and frigates as a Marine Engineering Officer and Weapons Engineering Officer for seven years. Christian looks back on more than 40 years of involvement in the international maritime industry, having gained his management and project development experience through such positions as CEO of a Tanker and Offshore shipping company, MD in charge for the shipping activities of a leading finance syndicator for shipping investments, and the head of shipbuilding for a large German Industrial Group which included overseas postings. Germanischer Lloyd's Chief Commercial Officer at the time of the merger with DNV, he is now DNV's Naval Business Segment Director. He holds a degree in Mechanical Engineering of the Helmut Schmidt University of the Federal German Armed Forces, Hamburg and participated in an International Business Post-Graduate program at the University of Houston.

SUBMARINE VULNERABILITY REDUCTION: WYVERN® AND PURPLE FIRE

C S Ritchie and **J S Schofield**, Survivability Consulting Limited, UK, **A E Walchester**, and **D Patten**, BMT, UK

SUMMARY

Vulnerability reduction and management has become standard practice for UK warship designs, but the same processes have not been fully applied to submarine programmes. Reducing the vulnerability (chance of damage given a hit) of submarines has historically been of lower priority in the design process, with low susceptibility (low chance of taking a hit) being prioritised. However, there is always a risk of taking a hit, and increased weapons and technology proliferation may change this balance further. There are features that can be incorporated into submarine design that can significantly reduce vulnerability, as has been demonstrated in previous work for UK MOD.

This paper demonstrates the potential for vulnerability reduction by assessing the impact that various design features have on the vulnerability of BMT's WYVERN SSK design using SCL's Purple Fire Vulnerability and Lethality tool. The paper therefore seeks to advance the conversation on how wider submarine survivability should be approached.

NOMENCLATURE

AIP	Air Independent Propulsion
AUV	Autonomous Underwater Vehicle
ConOps	Concept of Operations
ESM	Electronic Support Measures
PECD	Probability of Escaping Critical Damage
RoB	Reserve of Buoyancy
SCL	Survivability Consulting Limited
SSK	Conventional diesel-electric patrol submarine

1. INTRODUCTION

1.1 NAVAL SURVIVABILITY

The survivability of a warship or submarine is a measure of the vessel's ability to carry out its designated missions in a combat environment and can be broken down into three constituent components [1]:

- Susceptibility covers the aspects that determine the probability of being hit, including detection, classification, targeting etc.;
- Vulnerability is the probability that following a hit some functionality is lost, immediately or at some short time after the hit; and
- Recoverability is the probability of this functionality being restored so that a mission can continue longer-term.

1.2 SURVIVABILITY IN DESIGN

A combination of features relating to these three dictates the overall survivability of a vessel. Submarines are by their nature difficult to detect and much of the focus of the last 100 years of submarine design has been on minimising their detectability, and hence susceptibility. Reducing acoustic, magnetic and other signatures has been a key consideration for designers in the UK [2] and US [3] in particular. This has perhaps contributed to the general belief that submarines cannot be built to survive a weapon hit. Russian design practice [4] also focuses on minimising susceptibility but arguably places more emphasis on reducing vulnerability than 'western' designs, as demonstrated by increased internal subdivision and Reserve of Buoyancy (RoB).

Design in general, and more specifically submarine design, is about managing compromise and balancing the competing requirements to produce a functional output. The concept of balance is of great importance for submarines and refers to the need to maintain a hydrostatically sound design whilst also providing the requisite key design elements, such as space, power, and cooling capacity. Hence, the aim of submarine design is to produce a balanced boat that also has an operational advantage over potential opponents, and part of that advantage includes the relative survivability of the submarine.

For all three areas of survivability, addressing issues which might be exploited by threats relevant to the Concept of Operations (ConOps) early in the design process is the easiest way to mitigate against them and this is why having an efficient and effective vulnerability management process in place at the beginning is key. This has become the standard practice for UK warship procurement processes, resulting in good, low vulnerability designs [5]. Vulnerability assessment is now a tried, tested, and indeed mandated process which, if applied to submarine procurement and design processes, could result in designs with both low susceptibility and low vulnerability, greatly improving their overall survivability. Such vulnerability reduction for a submarine has been demonstrated in previous work for UK MOD.

1.2 PROBLEM DEFINITION

As mentioned above, in many countries efforts to improve survivability in submarine design focus on minimising the susceptibility, not necessarily vulnerability or recoverability. Vulnerability reduction for submarines has primarily focussed on mitigating the effects of underwater shock and whipping [2], by considering specific equipment shock tolerance and whole-boat structure. Although significant effort can be expended on these specific aspects of the design, doing so does not cover the whole picture. Further reducing vulnerability and hence improving the overall survivability would reduce the risk of mission failure, loss of life and loss of the submarine in a combat scenario.

Therefore, BMT and SCL have embarked on a joint study to identify novel design features intended to reduce submarine vulnerability. This study is limited to investigating the performance of design features and has not attempted to assess the impact of their implementation on the design. However, understanding the performance allows more-informed decisions to be made during the design process and identifies the features worthy of further investigation.

BMT produced the WYVERN® SSK concept design in 2017 to investigate the feasibility and capabilities of a small SSK [6]. It is briefly introduced in Section 2.

Purple Fire is a fast-running vulnerability and lethality tool developed by SCL and authorised by UK MOD. It has been used extensively across numerous warship assessments since 2015. It was introduced in a previous RINA paper [1] so detailed treatment is not given here, but the vulnerability management process is covered in Section 3 and the WYVERN Purple Fire model is shown in Section 4.

The paper then uses the WYVERN concept design, and applies design features to it with the aim of reducing its vulnerability. This results in a number of models within Purple Fire, assessed against a range of threats that the design could realistically expect to encounter given its ConOps, the results from which show the potential reductions in vulnerability achievable from each design feature.

2. WYVERN SSK

The submarine design (as shown in Figure 1) has a submerged displacement of 860 tonnes, 4 torpedo tubes and an Air Independent Propulsion (AIP) system. Since its inception, the design has been used as a testbed to investigate various capabilities which has resulted in several design variants, including the integration of an Autonomous Underwater Vehicle (AUV) hull plug [7], external batteries [8] and a variant optimised for a Coastguard role [9].

Figure 1: BMT WYVERN® SSK Design [6]

WYVERN was designed with simplicity in mind and is intended to have the minimum capability required to fulfil the ConOps and provide a 'credible threat'. Its small size makes it ideal for littoral operations in shallow and congested environments. The design's survivability is ensured by its submerged endurance, which is enhanced by the AIP system, coupled with a comprehensive sonar suite to give it very low susceptibility. The design also makes use of a double hull, primarily for arrangement purposes, which increases weapon standoff and improves vulnerability. However, the littoral environment means that this may not always be enough, with threats such as mines, ever quieter submarines and airborne assets being of concern. Therefore, it is desirable to identify features which could be incorporated into future design iterations to reduce the vulnerability.

3. THE VULNERABILITY MANAGEMENT PROCESS

The general vulnerability design process consists of three primary areas, each of which feeds into each other.

3.1 VULNERABILITY REQUIREMENTS

To ensure that a design has a low vulnerability, it is necessary for numerical requirements to be set, and measured against. These requirements relate to the average probability that a capability is available after weapon damage has occurred, and are termed the Probability of Escaping Critical Damage (PECD) for each capability. The numerical values of the PECD requirements are typically set for traditional "Float", "Move", and "Fight" capabilities of a vessel versus given threats appropriate to the design and its role.

3.2 VULNERABILITY MANAGEMENT

Typically, during the design process of a surface ship several rounds of vulnerability assessment take place as the design matures. This is a result of new general arrangements, changes to system design, or better maturity and understanding of equipment location and capabilities. These changes can significantly impact vulnerability and so it is crucial that these assessments take place early and often during the process in order to identify potential issues and mitigate against them. The assessment process easily keeps pace with a developing design at all stages from concept to service.

The UK MOD's Vulnerability Reduction Strategy is a structured approach that should be applied during the process, these are:

- Prevent catastrophic loss;
- Remove single points of failure;
- Concentrate remaining critical elements;
- Separate alternative sources of capability;
- Protect remaining vulnerabilities;
- Don't place critical items in areas most likely to be hit.

The above measures are not complicated, the aim is that they act as a simple method to focus the design of the warship. Whilst there are considerable differences between warship and submarine design considerations, there is no reason that the same management process and simple steps could not equally be applied during submarine design. The specific measures applied to improve submarine design will obviously differ to those applied to surface ships, with different areas of focus, however this should not be reason to not do so.

3.3 VULNERABILITY ASSESSMENT

The first step in performing a vulnerability assessment is to create the representation of the design, both physical and functional, within the modelling software, in this case Purple Fire. This requires information on:

- Hull and superstructure for a ship, or pressure hull and external structure for a submarine;
- Internal decks and bulkheads;
- Stiffening;
- Equipment required for Float, Move, and Fight functions;
- Cabling and pipework for systems;
- Logical representation of the above functions in order to create fault trees.

After this model has been created, the next step is to assess the model against the threats outlined within the requirements. This is usually a range of weapon types, each of which will have a specified set of points at which it is assessed. From this the analyst will know if the design has met its requirements, and if not, suggest measures which would improve the vulnerability performance such that they are met.

Routinely repeating this assessment throughout the design process, including the assessment of potential design options at each stage, allows for continual monitoring of PECD performance as the design changes physically or functionally.

4. PURPLE FIRE WYVERN MODELS

4.1 BASELINE MODEL

The first step in this investigation was to create the baseline WYVERN model within Purple Fire. By using a 3D CAD representation, the majority of the structural elements were imported directly. The CAD file also allowed for accurate placement and sizing of the equipment. As WYVERN is a concept design, the Purple Fire model shown in Figure 2 with the external structure and equipment shown, is sparser in equipment detail than a final vulnerability model would be after the detailed design phase.

Figure 2: WYVERN Purple Fire model, outer casing and equipment

As well as the casing and equipment, a significant amount of structure exists between the pressure hull and outer casing within the model. This is a very important part of the modelling of submarines, particularly for double-hulled designs, as this structure will interact with any shaped charge penetrator, and degrade its penetrative capability before the important pressure hull hit. The inter-hull structure as well as the internal decks and bulkheads are shown in Figure 3.

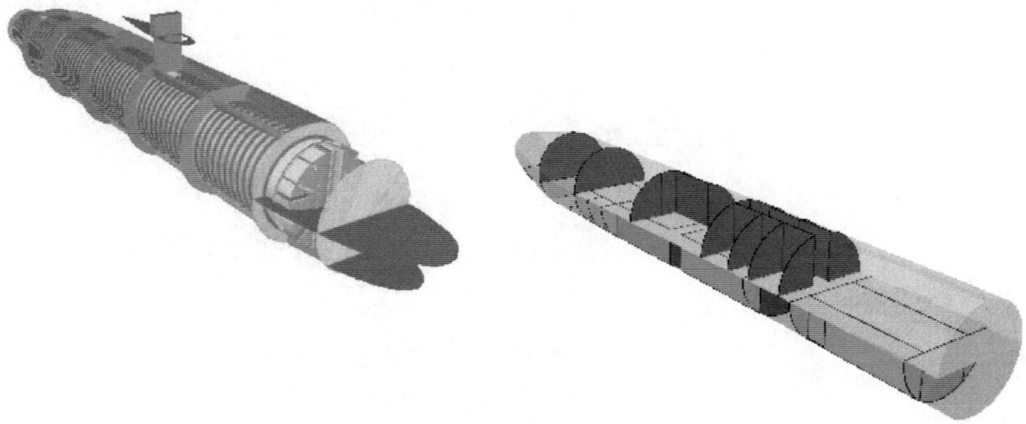

Figure 3: WYVERN Purple Fire model, inter-hull structure and internal decks & bulkheads

Even though the model is relatively basic in its concept form, there are still significant amounts of equipment modelled, which are then implemented into functional systems. These top-level systems are summarised below in Table 1. The ConOps also identified two missions that WYVERN may be required to perform. These are also included in Table 1.

Table 1. WYVERN Top Level Systems/Missions

Top Level System/Mission	Description
Surface and Abandonment	Depth control and access outside boat
Minimum Mobility	AIP, Cruise, or Main Motor Propulsion & Steering Control
Cruise Mobility	Cruise, or Main Motor Propulsion & Steering Control
Maximum Mobility	Main Motor Propulsion & Steering Control
Min Threat Detection (UW)	Bow, Towed, or Flank Array available
Min Threat Detection (Surfaced)	Electronic Support Measures, Optronics Mast, or Radar
Min Threat Avoidance	Cruise Mobility, Threat Detection (UW & Surfaced), Countermeasure Launch
Min Torpedo Launch	Minimum Mobility, Threat Detection, Torpedo Fire
Full Fight	Maximum Mobility, Threat Avoidance, Torpedo Fire
Mission 1: Protection ASW	Minimum Mobility, Threat Detection (UW), Torpedo Fire
Mission 2: Credible Threat	Minimum Mobility, Threat Detection (UW & Surfaced), Torpedo Fire

For the vulnerability assessment, these systems are implemented as fault trees within Purple Fire. An example of one of the top-level systems is shown in Figure 4, alongside the key to the diagram. Simply put, following a damage event, the vulnerability of the system is determined by the ability to trace a path from top to bottom of the fault tree via a route with no damaged equipment; if such a route exists the system is not vulnerable. In reality, the assessment is probabilistic and a Monte Carlo calculation is used.

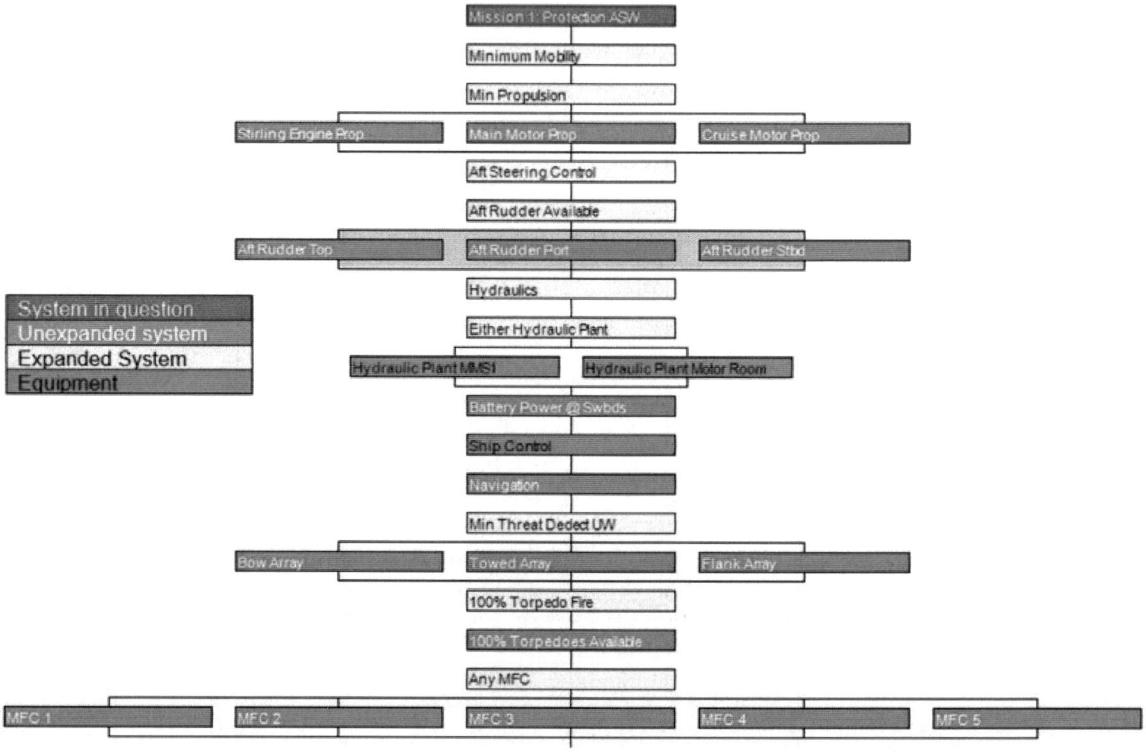

Figure 4: Example of Top-Level fault tree

4.2 REDUCED VULNERABILITY DESIGN FEATURES

As outlined previously, several design features that could change vulnerability were then implemented into the model. These are summarised in Table 2.

Table 2. WYVERN Reduced Vulnerability Design Features

Feature	Description
1	Removal of the existing double hull structure
2	Shock mounting the torpedoes within the tubes
3	Shaft Independent Propulsion System (SIPS) & separation of key items of equipment
4	Increased shock capabilities for "Fight" and "Move" systems
5	Shaped charge armour protection applied to pressure hull
6	Alternative structural materials and reduced frame spacing
7	Enhancing external structure and control surface shock capability
8	Create three watertight subdivisions, with flooding of any one survivable
4&8	Features 4 & 8 combined
5&8	Features 5 & 8 combined
2-8	Features 2-8 combined

Feature 1: Removal of double hull
As WYVERN already has a double hull (i.e. outer casing surrounds the whole pressure hull), Feature 1 was designed to show the vulnerability benefit which already exists.

Feature 2: Shock mounting torpedoes
The aim was to determine the system level vulnerability due to torpedoes being stored within their tubes, and the reduction in vulnerability possible by shock mounting them (which is possible given the design is for torpedoes which "swim-out" of the tubes).

Feature 3: SIPS with separation
The introduction of the SIPS and the separation of key, duplicated items of equipment (e.g. switchboards) should improve the vulnerability of minimum propulsion.

Feature 4: Increased Fight & Move shock capability
The shock resilience of all equipment necessary for "Fight" and "Move" equipment were increased towards that of the "Float" equipment.

Feature 5: Shaped Charge protection
A layer of material was added to the pressure hull with sufficient thickness to stop a significant shaped charge penetrating into the pressure hull.

Feature 6: Structural Enhancement
The material properties of the outer casing and pressure hull were changed (the outer casing to a thicker titanium, the pressure hull to thicker steel with a higher yield strength), and the frame spacing reduced.

Feature 7: Control Surface Enhancement
The keel material was changed to titanium and its thickness increased, the sections of outer casing constructed of composite changed to steel, and the shock failure thresholds of the control surface equipment increased.

Feature 8: Watertight Subdivision
Two watertight bulkheads were added, splitting the boat into three, roughly equal in length, watertight zones. For the purpose of showing the benefits to vulnerability, it was assumed that the boat would not be lost if flooding occurred in a single zone.

Whilst it is noted that some of these features would require significant rework to result in a balanced submarine design, the aim of this assessment is to show the bounds of the potential improvements to the design's vulnerability. Future work would need to be undertaken to explore the viability of these features, and their impact on the overall design and cost. The key point to vulnerability reduction is that the biggest benefits are achieved when considered from pre-concept, when they can be implemented with least rework, and often using a combination of individual features, hence the combinations in Table 1. This process is what occurs in UK warship design.

5. WYVERN VULNERABILITY ASSESSMENT

Having included the features in versions of the Purple Fire model, the next step was to perform a vulnerability assessment against each. This would firstly determine the vulnerability of the baseline WYVERN model, and secondly determine how much of a reduction in this vulnerability each of these features or their combinations provides.

5.1 THREATS AND GRIDS

As previously described, vulnerability assessments assess a range of threats each considered at a series of potential points (referred to as the "grid") against the functions of the boat. This assessment assessed the vulnerability of the WYVERN baseline and variation with the above features against three weapon types and grids.

5.1 (a) Torpedo

The first of these was the assessment of a lightweight torpedo (LWT) and heavyweight torpedo (HWT) in a grid of uniformly distributed points around the outer-casing, as shown in Figure 5. The LWTs were considered including and excluding a shaped charge. From each of the points of the grid, the torpedo runs towards the submarine (along the hull normal at the point) and detonates at a standoff appropriate to the warhead position with a LWT or HWT.

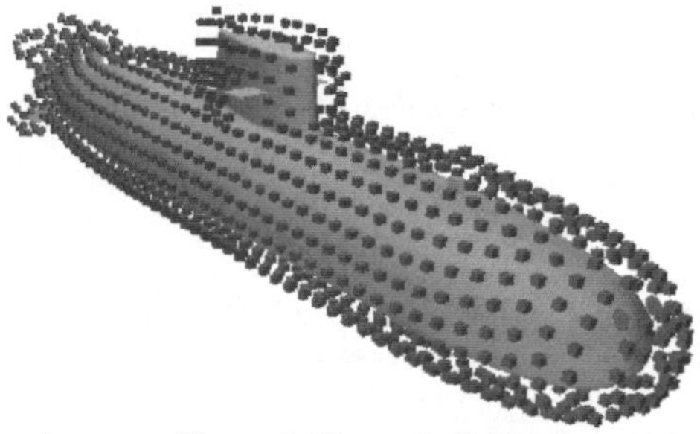

Figure 5: Torpedo Grid

5.1 (b) Depth Charge

The Depth Charge Grid shown in Figure 6 is a plane of attacks covering the boat itself and a given distance from its extents. The points represent a grid from above with the weapon fusing set so that detonations occur midway up the outer casing, except for those that contact the boat which detonate at that location.

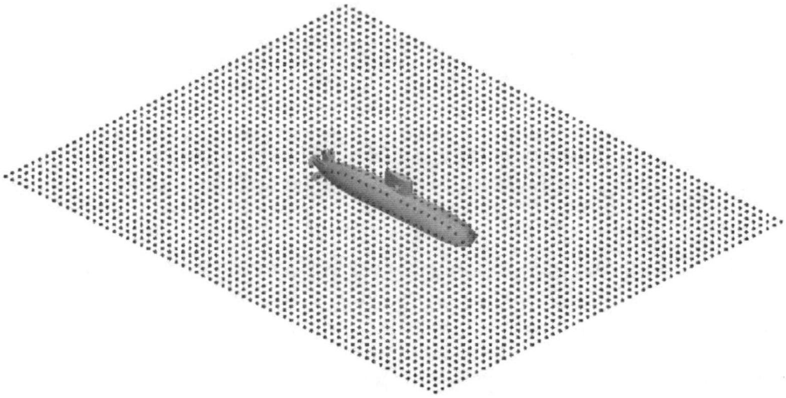

Figure 6: Depth Charge Grid

5.1 (c) Ground Mine

The Ground Mine grid shown in Figure 7 is a large volume grid underneath the submarine. The boundaries of this volume correspond to a distance which causes a minimum level of shock loading to WYVERN. Purple Fire automatically sets the seabed position to be coincident with the detonation position for ground mines.

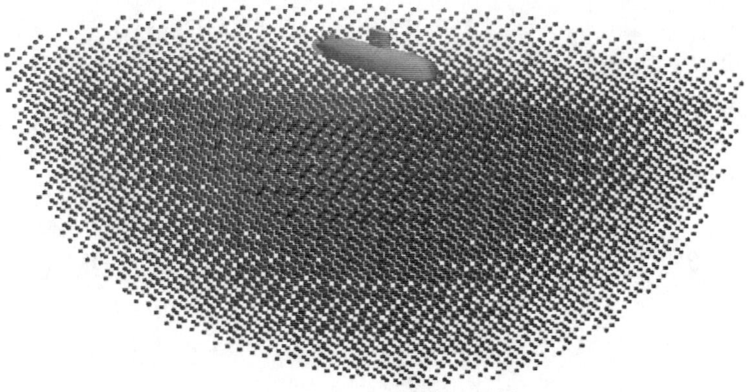

Figure 7: Ground Mine Grid

5.2 VULNERABILITY RESULTS

The Purple Fire assessment comprised simulation of the detonation of each threat at each point of its grid in turn, with the probability of kill for each system calculated at each point. From this, two key outputs are available for analysis; numerically the average of these probabilities over all points in a grid (the PECD) and visually the "vulnerability footprint" showing where a particular system is vulnerability to a particular threat.

After the thousands of different attack scenarios were considered against the Purple Fire models, it was clear that some of the design features resulted in significant reductions to WYVERN's vulnerability. Not only do some of the measures result in a reduction in system vulnerability (i.e. probability that a given capability is unavailable), some could also greatly reduce the likelihood of the total loss of the boat. Figure 8 below shows an example of these vulnerability reductions versus the Wyvern baseline vulnerability, for two top-level systems and total loss of the boat for each feature (or features) against a combination of some of the threats.

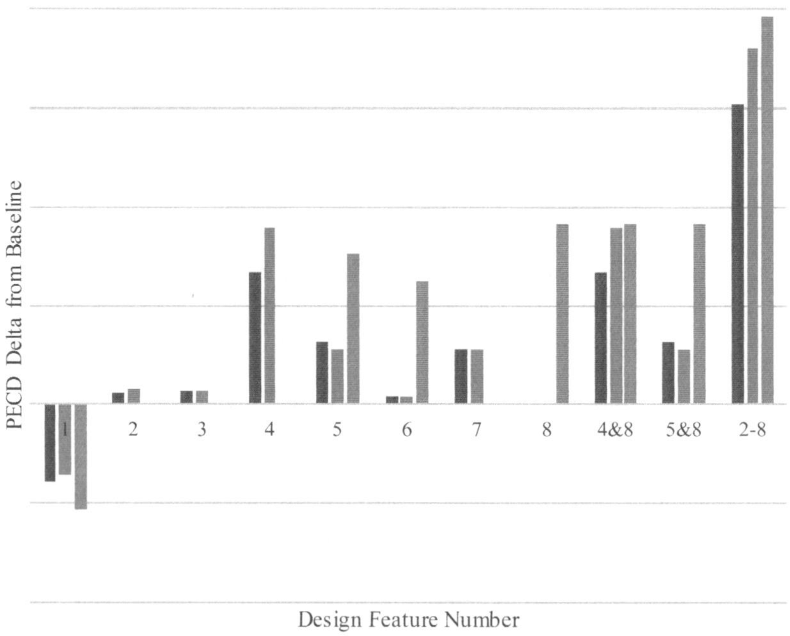

Figure 8: Example of potential PECD changes for two top-level systems (blue & orange) and total loss of the boat (grey) compared to the baseline WYVERN design

As seen in numerous surface ship vulnerability assessments, the greatest reductions in vulnerability occur once multiple issues have been solved, resulting in this case with the combination of all features outperforming the baseline model and any of the individual options on their own.

A wider report not suitable for this paper analyses the vulnerability of each feature in detail. However, a few of the options which resulted in the greatest benefit will be discussed here.

Changing WYVERN to be single-hulled, Feature 1, saw a significant increase in its vulnerability, highlighting the gains already present on the baseline design. With a breach of the pressure hull assumed to result in a non-survivable level of flooding, the threats detonating closer to the pressure hull in a single-hulled design can result in failure of the pressure hull or penetrations, and the loss of the boat. So, already incorporating this design feature in WYVERN has significant benefits.

Feature 3 introduced a SIPS with separated critical items of equipment. This is primarily based on lessons learned from surface ship vulnerability assessments, introducing separated redundancy. Whilst this is relatively "easy" for surface ships due to their size, it is obviously much more difficult to achieve on a submarine of WYVERN's size. However, whilst the gains were limited for WYVERN, the concept is still correct, and in a larger submarine this would provide significant benefit should redundancy be implemented with separation.

This feature raises an important point, that whilst the lessons learned during surface ship vulnerability designs can apply to submarines, the differences in effectiveness of some of these methods will vary as will their viability. However, it is important that these lessons are not forgotten simply because they may have to be applied differently.

Feature 4 saw the shock failure threshold of equipment required for "Fight" and "Move" functions increased. Whilst in reality this would be an extensive (but entirely achievable) undertaking the benefits to vulnerability are significant. With shock damage being a prominent mechanism in the underwater theatre, particularly from mines in the littoral region in which WYVERN could be expected to operate, mitigating against this is of obvious benefit to vulnerability. In this case it could offer WYVERN the ability to perform roles/missions that otherwise would have been deemed of excessive risk.

Feature 5, applying armour to the pressure hull, also results in significant improvements in vulnerability. However, these are obviously limited to specific threats. As with Feature 3, this raises an important point, that whilst there are many of these vulnerability reduction measures that result in significant gains, a driving factor will always be the customer's requirements. Should the customer expect to utilise WYVERN primarily against other SSKs then maybe protecting against shaped charge penetration would be of greater importance than other measures. Likewise, if it is more likely to be used in a littoral role for special forces insertion and reconnaissance then potentially improving the ability to withstand shock damage from ground mines is of greater importance. From the concept stage, changes could and would be made in line with a customer's requirements, from which the measures for reducing vulnerability will become more or less applicable.

Introducing watertight subdivision (Feature 8) can greatly reduce the probability of loss of the boat (Figure 8 grey). However, without more than one additional feature the vulnerability of the top-level systems to some of the threats, including those contributing to the example results in Figure 8, are not greatly improved by Feature 8 in isolation (top-level systems blue and orange). WYVERN is likely too small to benefit across the threat spectrum from this feature, although it has previously been seen to be more effective on larger boats.

Combining Features 4 and 8 resulted in a design that was both very survivable to shock damage, and might remain afloat with some level of flooding within the pressure hull. The vulnerability reduction produced by Feature 4 has already been discussed. The benefit of Feature 8, having a survivable level of pressure hull flooding, is the reduction in the potential loss of the boat against some threats. The vulnerability of some top-level systems has been reduced significantly, and the likelihood of the boat's loss versus some threats has been almost totally removed. Introducing measures such as separating redundant equipment or armouring the pressure hull could provide further vulnerability reductions. The reduction in vulnerability with Features 4 and 8 is shown below in Figure 9, which shows an example of a system's vulnerability footprint compared to the baseline, with red points being locations with a high probability of kill for the system in question.

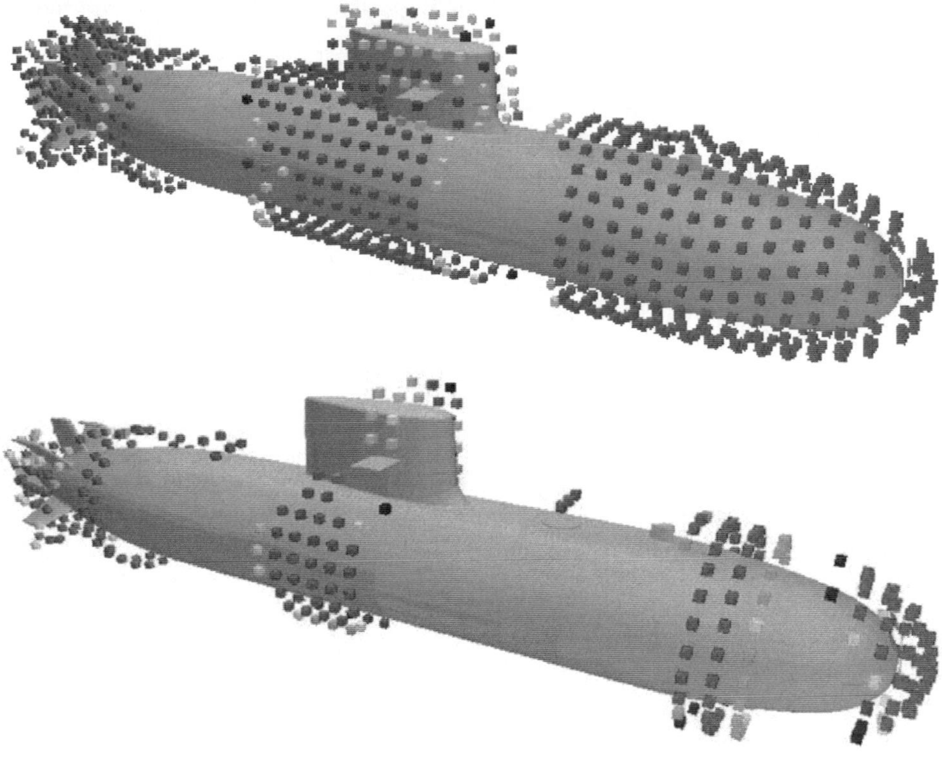

Figure 9: Baseline (top) vs Features 4 and 8 (bottom) vulnerability footprint for an example system

6. CONCLUSIONS

Submarine design will always be a managed compromise, but improved understanding of the benefits and performance of various vulnerability reduction methods allows for better informed design decisions. This paper has shown that significant reductions in vulnerability are possible to achieve through the implementation of some of these design features. The successful implementation of these features in a balanced submarine design presents the next challenge to be overcome, and it is expected that achieving a balanced design without unacceptable sacrifices to other elements of the design will present the greatest challenge.

Importantly, the vulnerability lessons learned during previous surface ship design processes are generally applicable to the vulnerability of submarines. Whilst again acknowledging that some of these features would be difficult to implement, it is clear that there is a worthwhile cost/benefit debate to be had regarding submarine vulnerability. Would an enemy combatant engage WYVERN first if it was known to be a survivable design which required accurate targeting to defeat? And would a more survivable design allow WYVERN to perform missions that would have otherwise been deemed too risky? That is not for this study to determine, but hopefully will encourage future debate on the matter of submarine vulnerability.

8. REFERENCES

1. SCHOFIELD, J S., 'Purple Fire Vulnerability/Lethality Assessment Tool', *Transactions of RINA*, Sept 2018.

2. BURCHER, R., RYDILL, L., 'Concepts in Submarine Design', *Cambridge University Press,* 1994

3. JACKSON, Capt. H A., 'Fundamentals of Submarine Concept Design', *SNAME Transactions, Vol. 100*, 1992.

4. KORMILITSIN, K N., KHALIZEV, O A., 'Theory of Submarine Design', *Riviera Maritime Media*, 2001

5. SCHOFIELD, J S., et al., 'The Influence of Survivability in the Design of a Capable, Affordable Surface Combatant', *Transactions of RINA*, June 2012.

6. BMT, 'WYVERN' https://www.bmt.org/projects/project/3322/wyvern (Accessed 29th April 2023).

7. MORRIS, D., BUCKTON, P., 'The Design Challenges of a Submarine Host AUV Hull-Plug', *Undersea Defence Technology 2018*, 2018.

8. AINSWORTH, W., 'Benefits and Integration Aspects of External Submarine Batteries', *Undersea Defence Technology 2018*, 2018.

9. WALCHESTER, A E., 'The Coastguard Submarine: Covert Maritime Security', *Undersea Defence Technology 2019*, 2019.

9. AUTHORS' BIOGRAPHY

Crawford Ritchie, MEng, holds the position of Senior Survivability Engineer at Survivability Consulting Limited. He is responsible for performing vulnerability and lethality assessments. His previous experience includes having undertaken vulnerability assessments of Type 26 and its export variants, Type 31, the Fleet Solid Support ship and assessments across the in-service fleet, as well as assessing the lethality of current and future torpedo programmes

Alex Walchester, CEng MEng MRINA, holds the position of Senior Naval Architect at BMT. He is responsible for undertaking naval architecture work, primarily focussed on submarines. His previous experience includes tasks ranging from concept design to in-service design change. Previously he has presented on the use of reference designs and using submarines for maritime security.

Dan Patten CEng MSc (Hons) MRINA is a Senior Naval Architect at BMT, with 13 years experience as a Naval Architect specialising in Survivability. He is currently Survivability Manager for the Fleet Solid Support Ships project, having recently returned from secondment to the Type 31 project, where he was the lead survivability engineer.

James Schofield, CMath MMath MA (Cantab) MRINA FIMA, is Managing Director of Survivability Consulting Limited. He is the lead developer of Purple Fire and has 21 years of experience in naval vulnerability and lethality assessment, advice and policy.

CAPABILITY VERSUS AVAILABILITY IN SUBMARINE DESIGN?

Professor D Andrews, University College London

SUMMARY

Capability and Availability are seen as terms adopted to encapsulate major features required in designing a new major warship, such as a submarine. With these terms it is worth considering what they mean when it comes to designing such complex naval vessels. This paper firstly considers both of these terms in turn and questions whether capability is often too narrowly defined and, aside from affordability issues, which are clearly crucial in designing any complex system to meet a set of capabilities, capability could be seen to be all embracing in regard to the set of qualities desired for a new class of vessels. If this is so, then "capability" ought to cover availability among other whole boat performance characteristics.

Turning to availability, from a design stance it is considered this has been something of a Cinderella topic when scrutinising discussions on submarine design in the open literature. Thus, there follows a brief review of some key published sources, including those presented at the RINA submarine conferences, which started with the RINA 1983 International Submarine Symposium. This leads to consideration as to why availability seems to be addressed relatively late in the traditional submarine design process, such that any drop off in a class of submarines, in regard to that force's expected availability, might not be revealed until such a new squadron of vessels is fully in-service. This then raises the question as to how the process of designing such complex vessels might be better focused, in early stage design, to address the issue of availability. Without better consideration of fleet wide availability, the other operational capabilities sought in an affordable new design are unlikely to be effectively deployed at sea to a cost-effective level of availability. This new focus for submarine designers, is seen to be best addressed from the start by adopting an architecturally based "ship synthesis" approach and to be assessed at a fleet or class level.

1. INTRODUCTION – CAPABILITY VERSUS AVAILABILITY - REALLY?

"...the readiness is all." William Shakespeare (HAMLET)

Capability is an oft used term in the early consideration of a new naval vessel. Its usage can be traced back to US Defense Department's practice post-WWII (Hitchens, 1986), which was driven by a desire to change the emphasis in evolving new military equipment, from performance focused specific needs, to emphasise instead the military capability that the new equipment is intended to meet. Whether this was a good change in procurement practice for naval vessel acquisition is open to question, as this author has recently debated to a USA audience (see Andrews, 2022) when addressing a systems approach applied to naval ship design.

If we consider "capability", does it just mean the combat system part of the whole warship? This is often how it is seen by naval combat engineers, and even the more generally focused systems engineers, the latter of whom ought to know better (see again Andrews (2022) demolition of such engineers' usage, summarised in the next section, of the terms platform and payload). If "capability' is intended to denote what the new warship provides to the fleet, leaving aside all that "Float", Move" and "Infrastructure" provide to the "Fight" (combat system) aspects of capability (see Andrews et al, 1996 for definitions of the above terms defining a warship's functions), then capability must also include, for example "stealth", which itself permeates through out a submarine design in a "transversal" manner. But if submarine mobility (i.e. Move) is taken to include the need to dive to a specific depth and all that is implied by "safety considerations", for such a safety sensitive vehicle as a fast deep diving submarine, then there can be little in submarine design that is not a necessary component of capability. This then leaves little regarding the performance of a new design that can be sensibly omitted from "capability" beyond, perhaps, "affordability". Given this last aspect addresses the price to be paid for a set of capabilities, it can be seen as the other prime design consideration, alongside performance/capability, in achieving a "cost-effective" solution? So, should such a wide consideration of capability, balanced by affordability, not also include availability or, as we shall see, "fleet readiness profile"?

It is worth considering why availability has been something of a Cinderella consideration in submarine design (and indeed warship design in general) and, by the implication of this conference's title, whether it should be seen as separate from capability, which this paper's opening paragraphs have argued should not be narrowly (combat system) defined but include all design aspects, including availability, but perhaps not affordability, when adopting a cost-effective acquisition stance?

In two recent papers to previous submarine conferences (Andrews 2017, 2021), the author has argued that availability is a significant capability to be sought in submarine design, alongside all the many other interacting capabilities (well beyond combat system features) that the submarine designer seeks to provide in any new submarine design. Thus, in the 2017 paper outlining the unique set of issues associated with achieving a balanced submarine design, integrated logistic support (ILS) considerations were said to be a significant issue to address early in submarine design, where availability ought to have been taken to be included among the ILS suite of issues. And when the set of typical design choices (which need to be coupled with meeting the demanding physics involved in operating a submarine at speed and depth) are sequentially listed in the second (2021) paper, the order in which they are discussed is clearly qualified by the fact that these choices really are all interrelated and must be considered as a whole. One of the clear considerations (capabilities?) is delivery of a high level of availability – and that means, as will be seen below, the availability achieved should be that for the whole fleet of submarines that a navy wishes to deploy.

Availability should be easy to define: most engineers are introduced to the triumvirate of ARM (Availability, Reliability and Maintainability) early in their education and then get on with the things that interest them. A recent major focus on Availability in submarines, to which we will return later, highlighted the lack of precisely that aspect when the Collins Class SSKs were introduced into service in the Australian Navy. Fortuitously, the Commonwealth of Australia has published several detailed reports on the work of the team led by John Coles, into this severe drop-off. The second Coles Report (Coles, 2012) adopts the term "Sustainability" rather than availability but spells out, for the six Collins Class vessels, the availability of that fleet in terms of Material Ready Days (MRD) benchmarked to an "Average performance of international programs" using that term scored at 1.0 against the RAN achievement of only 50%. This, without the (classified) actual MRDs, is a stark comparison and puts in poor value for money (VFM) terms any focus on "capability" without spelling out what Coles identifies as "a readiness and availability requirement for the class".

Thus, the current paper next considers how (if at all) the classic texts on submarine design, including many in previous RINA submarine conferences, since the first in 1983, have addressed availability and, importantly, in the design phases, how an acceptable "capability" in that regard might be achieved. Section 3 considers, from perusal of many of the specialist texts on submarine design, why, given its importance, availability would seem to be relatively rarely addressed in the vital early design stages. Section 4 looks at different strategies for designing and acquiring submarine and how they place availability in context. Section 5 before the concluding remarks, considers if and how in the overall submarine design process, the issue of submarine readiness and availability can be better addressed by being regarded as a distinct focus within the overall capability sought. In particular, this is seen as changing the mind set to the sustainability of a given fleet of submarines, rather than being focused on an individual submarine's capability.

2. HISTORIC CONSIDERATION OF AVAILABILITY IN SUBMARINE DESIGN

In considering how availability has been addressed in submarine design practice, one could start with the classic textbook produced by Burcher & Rydill (1994) and which is still the basis for the only open post-Masters level short course in submarine design. This has been run at UCL since the early 1970s (www.ucl.ac.uk/mechanical-engineering/study/other-programmes/submarine-design-acquistion-course). With their focus on early-stage design, the term "availability" does not even figure in the book's Index, 'though there is throughout the book a strong emphasis on safety, which is a significant driving concern in maintaining submarine availability.

Early papers on modern submarine design in the RINA submarine conferences, from the first in 1983, have also focused largely on direct design issues, with only a few papers addressing thought life support in general and even fewer directly considering availability.

At the 1983 RINA Submarine Symposium, Daniel's scene setting paper "Considerations influencing Submarine Design", from a major player in UK early SSN/SSBN design programmes prior to his being in charge of the, then, nationalised UK warship shipbuilders, gives an overview of the main submarine design issues. Daniel's (1983) exposition culminates with the introduction of the UK SWIFTSURE Class SSNs. With this class, the nuclear attack submarine attained a design maturity, as has previously been outlined by the current author in his papers to recent RINA submarine conferences mentioned in the Introduction (Andrews, 2017, 2021). Apart from additional sections, beyond the usual naval architectural submarine topics, where Daniel discussed the sensitive issues of Survivability and Noise, he gives a submarine designer's view of machinery and life support. Thus, Daniel overlooked the whole set of issues concerning through life support, including the need for an extensive investment in nuclear power related shore support facilities as part of maintaining the availability of the fleet as a whole. This was common in other "design focused' early submarine conference papers, as is summarised in the brief reviews of a range of seminal papers below.

An even earlier review paper in the SNAME Transactions (Arentzen & Mandel, 1960) provided a very comprehensive (over 50 pages) state of the art on the "Naval Architectural Aspects of Submarine Design" from a US Navy perspective. In the paper and its extensive written discussion, the naval architectural aspects were (interestingly) limited to what could be considered the "classical" view of naval architecture in "ship design" by focusing on the S^4 ship design aspects of Stability, Speed, Strength, and Seakeeping (see Brown & Andrews, 1980). It should be noticed that the 5th "ship design" aspect, that of Style, was not explicitly addressed, perhaps being seen as more of a term to cover broader design issues than the other four traditional ship design concerns. Thus, of the wider issues relevant to the submarine design process, in part addressed in Burcher & Rydill's textbook (1994), Arentzen and Mandel just deal with build cost, such that through life supportability, and hence availability, are implicitly not seen as directly part of the naval architecture of submarines.

An excellent detailed presentation of the UK design of a modern SSK (now operated by the RCN) was presented in the RINA Transactions by Wrobel (1985). It provides an exposition on that specific design, and when it comes to issues associated with through life support and availability, Wrobel draws attention to the incorporation of a "Dutch Breech" in the pressure hull to facilitate removal of the main diesel engines. This was a significant design investment to improve availability. A similar feature in the UK Trident VANGUARD Class, taken from the USN OHIO SSBNs, was to incorporate two Logistics and Escape Towers (LETs), each giving a 6 foot diameter opening in the pressure hull with appropriate in-board removal routes to facilitate the removal of much of the boat's equipment.

A good review by Prins and Everard (in 1996 RINA Submarine conference) presented a Dutch/VSEL (now BAES Barrow) perspective of 15 conventional submarine designs, each with a cutaway profile focused on propulsion features. An interesting view (Fig 4) of "design iterations" covers ownership and support issues, however the term performance is not seen as needing to address availability. However, Prins and Everard do cover reliability and redundancy, and both aspects could be said to contribute to availability but here they only consider both issues with respect to Survivability and in considering the implementation of several AIP options, rather than its role regarding the whole set of submarine design choices (Andrews, 2021).

Prins and Everard, (1996) like many presenters, use the term "platform" and so it is considered worth, for this specific submarine audience, to comment as to why this is so inappropriate for something as complex as a military submarine and, specifically, nuclear powered attack SSNs. It would seem this usage originated with naval combat engineers (and their close engineering colleagues, systems engineers), who have gone seriously wrong and practiced a poor systems approach by applying the term 'platform' to naval vessels. In contrast certain software design engineers devised the term systems architecture, and in doing so say that they have followed the practice of naval architects (Maier, 1998)), and thus the appropriateness of systems architecture compared with 'classical' systems engineering has been outlined for complex vessels by Andrews (2022). While a naval combatant (or a submarine) can provide a physical platform for complex vehicles, like helicopters and UUVs, it is wrong to use the term "Platform" to describe a naval vessel as such. It is a whole system of systems and the further split into 'payload' and 'platform' (i.e. combat system and the 'rest of the ship') is a very bad design and acquisition mind-set. It implies the former is good and must be maximised at the expense of the latter.

So, the so-called "platform half" of 'the platform' (itself a tautological nonsense) provides flotation/survivability and mobility, both of essential military worth. That 'ship half' (which it may be in cost of acquisition terms, but is far more than 50% in terms of whole boat space demand, that directly drives ship and boat size) also provides the infrastructure for the personnel (who are largely 'fighting' the vessel) as well as the supporting ship services (such as electrics and chilled water) without which the combat system elements could not function at all. All of which shows the utter nonsense of the "two halves", which amounts to not seeing the warship as a whole system of systems. Thus, all the interdependent functions contribute to the vessel's capability, which means there is no unnecessary 'platform overhead' to be minimised. This typical combat system and procurement policy mind-set has led to false justifications to reduce whole ship capabilities in vital features, such as those contributing to sustainability, supportability, adaptability, robustness and mission flexibility, with through life cost consequences well beyond individual designs. The example of the "margin-less and short life" Type 23 Frigate programme is discussed in Section 4 of Andrews (2018) and the consequence of the Type 42 destroyers 'platform' inadequacies in the Falklands Campaign have been outlined in some detail in a forensic discussion of the losses in that campaign in Andrews (2022).

A paper addressing the RN nuclear submarine "design challenge" was published in *Nuclear Energy* by a former DG Submarines (Betts, 1999). Betts rightly commences with the design process, seeing it as a process unique to nuclear submarines, with four clear categories needed in a "draft" set of requirements: role; operating environment; performance; and, directly, availability, before summarising an extensive concept design. Betts also reflects on the introduction of the UK Ministry of Defence design tool SUBCON (Andrews et al.,1996) as a graphically based CAD tool and then its subsequent object-oriented version in QinetiQ's Paramarine design suite (www.paramarine.qinetiq.com). This application of the UCL Design Building Block approach (Andrews & Pawling, 2003) has enabled early design consideration of accessibility for (among other aspects) maintenance envelops, in the congested layout of a submarine, to ease maintainability and directly contribute to ensuring adequate availability of a new class of vessels. Furthermore, Betts then addressed more than the usual naval architectural considerations in submarine design by including topics such as noise, safety and auxiliary systems, but omitted to emphasise availability issues. The section on construction mentions the, then, GEC Marconi (now BAES Barrow) intent to use both CADCAM and a modular build process, both of which ought to have emphasised too their joint advantage in exploiting the availability implications, but, yet again, this was not explicitly discussed.

In the same year in a keynote paper to the RINA submarine conference, Fuller (1996) stated in his paper's title that the "essentials of the submarine are: To dive; To move; To live and To survive". In calling for a "holistic" design approach, Fuller reviewed 126 RINA submarine conference papers to that date and while the largest number were those on design and powering/AIP, there were less than ten each on structures, life support, resistance & propulsion and on survival. So again, availability was noticeable by its absence. In the paper's concluding section, that author saw (post-Cold War) that, for the immediate future, (evolutionary) designs are likely to be no more than stop gaps "until *the future submarine* requirement *has been* articulated". He then listed six likely drivers of new design concepts, with the fifth being "low maintenance materials and equipments, easy hull access", which could be seen as a partial recognition of the true importance of availability in modern submarine design?

In that same conference Tasker and Wilcox (1999) (from Marconi Naval Systems (soon to be BAES, Submarines) and UK MoD, respectively) presented the RN ASTUTE Class SSN design, following its March 1997 prime contract placement, with (what proved to be wholly optimistic) first of class (FoC) in-service date in 2005. From the current conference's perspective, interestingly, that paper talked in outlining its design process of: "concurrent engineering"; a modular construction; and identified, in a final "Realisation" subsection, "Design for Support". This started with stating the "Astute Class is the first RN submarine designed for support from the outset", by emphasising the provision of "the first 8 boat years support within the contract price". Whether (after considerable time and cost renegotiation in the subsequent Astute contracts) even this limited through life support has been reflected in the "guaranteed" availability of the whole class, is clearly an open question. It is likely to remain so, at least in the public domain, until well beyond the current decade, when it might then be seen as just an historic issue with, hopefully, some lessons on enhancing availability being useful to future submarine programmes.

The first of several more recent American perspectives than the 1960 one above, was presented by Sullivan and Tibbitts in the naval submarine chapter of the SNAME two volumed "*Ship Design and Construction*" (Lamb, 2004). This comprehensive review starts with Missions; then submarine history; design drivers; features unique to submarines; and "ship" characteristics. So at least in talking of "Producibility Concerns and Modular Construction" in that last section, there was an opportunity to cover availability of a whole class of vessels but, as has been seen pretty consistently in this section of the paper, a focus on the many challenges in submarine design has all too often been at the level of an individual submarine's performance and production cost, rather than the proper focus on a navy's submarine force's "capability". Thus, it can be concluded from this brief review that Availability (like most ILS issues) needs to be far better considered in the "design fight", which has been largely between just capability (often combat system focused) and again, largely, Initial Cost of the individual FoC design rather than the fleet resource that a whole new class of submarines actually provides the modern navy.

3. WHY IS AVAILABILITY CONSIDERED SO LATE IN SUBMARINE DESIGN?

In order to answer the obvious question as to why availability has been largely disregarded by most publications on submarine design and seen (if at all) to be considered relatively late in the design process, along with the rest of ILS, it is instructive to explore those few papers, mainly presented at more recent submarine conferences, that have considered submarine design for support and whether they have specifically addressed availability.

A COLLINS Class paper from ASC and Saab Dynamics authors (Smith & Stenholm, 1996) addressed the system management (unfortunately using the term 'Platform Management', thus perpetuating that misnomer) when focusing on a higher degree of automation for, specifically, the manoeuvring and propulsion control needs as part of an integrated "Ships Control System" (SCS). It was realised that redundancy (always vital for all submarine systems with significant safety implications) needed to be addressed in the design of any SCS. In searching for higher levels of automation, the authors discussed "cost of the asset and its availability" and then pointedly stated direct crew costs (already a driver on levels of automation) "are usually small relative to costs associated with having a boat tied up alongside", yet go on to say this is "not readily assessed when suppliers are tending for a new design" and "a boat alongside *(inoperable or deployable?)* is the province of the politician", which seems bizarre as it is only the project manager and his engineering team that could actually address this. This degree of abrogation by part of the engineering design team seems pretty stark, and highly questionable, even before in this specific design, in-service subsequent revealed the failure to address the Collins Class availability. The last three sections of the paper addressed "Economic Considerations"; Operability"; and "Training" with, again, a focus on single boat's operability, rather than emphasising fleet availability, which would have been more relevant once several boats were in service and fleet availability was found to be severely deficient and poor Value for Money (VFM). (See the Coles Report (2012) for further consideration, discussed below.)

Nicod (2011) of DCNS provided to the 2011 RINA conference, a technically well-argued focus on maintenance issues in early ("Basic") submarine design with a refreshingly more ILS focused consideration, starting in ESD. Actual major design alternatives were explored for their performance versus cost for different "maintenance concepts". It is good to see at ESD the proposal that specific decisions could be made affecting availability, such as H.P. bottles maintenance and whether they are to be located outside or inside the pressure hull. Such decisions can have profound implications for the vessel's size and build cost versus ease of TL maintenance. This example is taken by the current author as a justification for adopting the UCL Design Building Block (DBB) approach to ship and submarine design synthesis (Andrews, 2018), however, one might argue with the performance index chosen by Nicod and the life cycle costing logic, using a "rhythm of maintenance". The latter was used to show the effect of whole boat configurational choices, which is to be applauded, 'though Coles investigation (2012) suggests that consideration at a fleet/class level would be a better metric of fleet ownership.

After two SSK focused papers, just discussed, Smith (2011) from Babcock International addressed an ILS driven approach to submarine design for, specifically, the UK Successor SSBN programme, where availability for naval forces deployment is transformed from direct naval warfare issues. Thus, the UK national priority to have a SSBN at sea at all times plus a standby SSBN available to go to sea at short notice, results in the traditional ARM consideration being up rated to "Mission Availability". Smith's design-for-support approach lists consideration of seven physical locations, for submarine support facilities: equipment onboard; at baseport; at other support facilities; and at the original equipment manufacturers. However, he commences with the issue of provision of adequate onboard access, defined as "the ability to access the system/equipment to carry out planned and unplanned support activities". This, I would contend, is further support for an architectural/'inside-out' (i.e. DBB (Andrews & Pawling, 2003)) initial synthesis approach to better achieve a support driven design with sufficiently practical access spaces being incorporated in the design before the submarine overall balance is fixed.

In 2012 a team led by John Coles, former Chief Executive of the UK Warship Support Agency, which had been the largest UK government agency by spend and, who was also Head of the RCNC, produced the second of several reports on "Sustaining Australia's Strategic Collins Class Submarine Capability" (Coles, 2012). Despite using the first of this conference's two key terms, that report addressed the second, namely Availability. The overall study by Coles' team was provoked by the evidence that the availability of the Collins Class SSK fleet was well below comparable SSK fleets. In his Introduction Coles lists his key observations, which included:

- "current level of availability..*is* ...poor" both absolutely and relatively;
- practically, for a six submarine fleet, three should "always be available" "and sometimes four" (see Figure 1);
- "logistics support arrangements" did not support that level of availability;
- "Reliability has been the poor relation" and that is "not simply a case of buying more spares…".

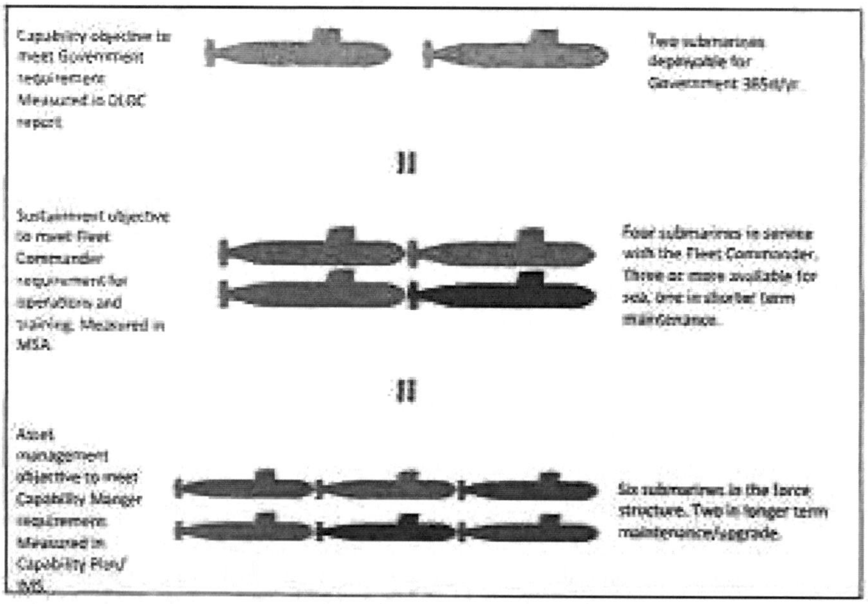

Figure 1: An example of a clear operational objective (Coles, 2012*)*
(Courtesy of BMT and Commonwealth of Australia)

The details in the Coles report justified the above criticism with five main issues identified and 25 Recommendations spelt out, if that fleet were to get to the level of 3, or even 4, boats availability. Most of Coles' recommendations were primarily, cultural and then organisational, fundamentally requiring the government and navy to be an intelligent customer for sustainability to achieve a level of availability for this fleet comparable to that of the best submarine fleets worldwide. From this report, which proscribed a retrospective remedial programme, it should be possible to read across to a future project for a fleet of SSKs (or even SSNs) what needs to be done to substantially enhance the acquisition programme's focus on fleet availability.

In the same time frame as the Coles' Study, a more direct US Navy "Design for Sustainment (DfS)" approach for the OHIO Replacement (OR) SSBN Class was published in the USA (Baker et al., 2015). It gave a very direct intent for this major programme to enhance the availability of the new fleet arising from a reduction in class size from 14 OHIO SSBNs to 12 OHIO Replacements, while still providing the same level of strategic deterrence. Thus, a Sustainment Engineering organisation has been set up at the shipyard (Electric Boat Corporation, Groton, CT) to design for the OR life cycle, and the paper outlined the manner in which this is being done. Baker et al. stated the SSBN design "has been directly and substantially improved" in regard to "material availability" through life and this is due to these DfS improvements being accomplished early enough. It would be interesting to see what this has meant in submarine design terms, but this is unlikely to be revealed for such sensitive vessels.

From the review of the above, more availability focused, publications, it is concluded its importance is becoming recognised and that the issue has to be addressed at a fleet/class level and, also, this has to be done right from the initial sizing/synthesis of a submarine design project. Furthermore, as Coles concludes, even for a class of SSKs, that they are expensive to maintain (even relative to surface warships) but the issue is not just one of providing extra money both in build, facilities and support resources (including crewing levels). More important is a Very High Level of national commitment and one can conclude, even for conventional submarines and yet more so for SSN/SSBN (see Macdonald & Nicholl (2022) discussed below), that tackling availability adequately imposes the need to manage a submarine design wholistically. This author considers this to mean that the "fuzzy" side of naval ship design and acquisition needs to be more directly addressed as part of managing the production of such complex designs (see Andrews, 2022), so the next section considers recent suggestions as to how this more holistic stance might be better achieved as part of a better focus on availability for a fleet of submarines.

4. MANAGING SUBMARINE DESIGN WHOLISTICALLY TO BETTER EMPHASISE AVAILABILITY

Systems engineering (S.E.) has (since the success of the Polaris acquisition programmes in the US and UK navies) been seen as synonymous with (naval) ship design, at least by some while being questioned by others (Rydill, 1966; van Griethuysen, 2000; Andrews, 2011). At the 2008 RINA Submarine conference, Gordon MacDonald of BMT, Australia, in talking of naval architects, systems engineering and submarines, showed (at his Figure 1) the "fundamental inputs to Capability" with support and facilities flagged up but not availability directly (MacDonald, 2008). The classic S.E. elements in the ubiquitous "Vee" diagram, as shown by MacDonald's illustration, lack feedback and the insight of the seminal Royal Academy of Engineering's publication (Elliott & Deasley, 2007). Their figure on page 16 of their report, which is reproduced below as Figure 1, shows in the first bubble, to obtain capability, requires cost, time scale and risk, all derivable only from "possible solutions". Furthermore, those most plausible solutions ought to explore "fleet availability" if they are to be consistent with this paper's message. MacDonald's paper goes on to compare the systems engineer at "the centre" as "key integrator" (for general projects see MacDonald's Figure 6) with the naval architect as the **ship** design key integrator (see MacDonald's Figure 7). One could use that author's "Maritime Systems Engineer" term but this author would rather name the naval architect as "(ship) systems architect" (Andrews, 2016). Either way, a holistic Through Life (TL) approach to ship design and ownership that addresses availability is clearly a significant part of the initial ship designer's (naval architect's) responsibility, even if this S.E example failed to address it directly, which is reinforced by the "Vee diagram" ending at acceptance, with a brief nod to "sustainability" and eventual disposal at end of life.

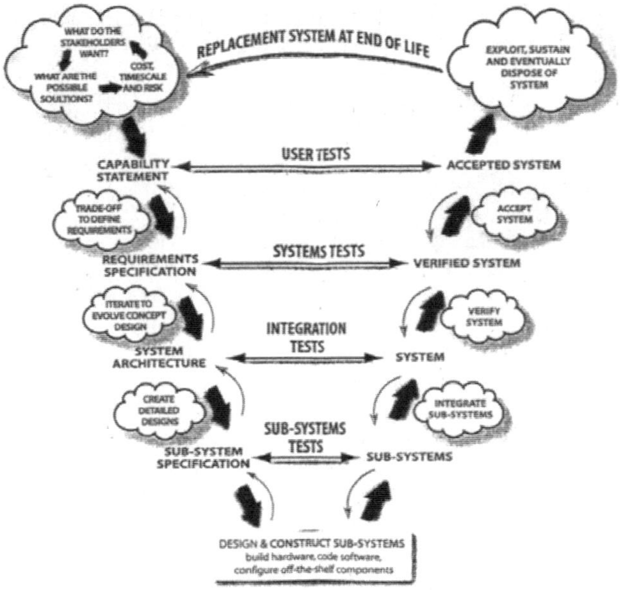

Figure 2: RAEng Vee Diagram for Complex Systems (Elliott & Deasley, 2007)

Another approach to naval ship design (with a SSK focus) could be said to have been presented by using Operational Analysis (OA) to derive the desired submarine capability for the Swedish Navy (Nordin, 2014). While Nordin purports to be able to make all the "design decision" using OA scenarios, the current author was able to question this regarding the classic omission in a SE approach when it comes to resorting to the old design chestnut "form follows function". As argued in Andrews (2018), this statement, due to the 19th Century architect Sullivan, is false (and has been recognised by architects, for many decades, to be a mere style choice but still accepted by most engineers). As Nordin shows in his Figure 21, Input Needs/Initial Requirements can only result from solution(s) (as in Figure 1) to be tested by the **designer** (as part of Requirement Elucidation) through the designer making choices on the Style of each individual solution when to synthesising new designs (Andrews, 2018)). Thus, Nordin designates "Set Design parameters and philosophy (Style)" (not to be confused with Set-Based Design - see next paragraph) to then size (using volume and weight distribution) and test for balance (for submarines this fundamentally consists of: weight/buoyancy; resistance/powering; vertical and longitudinal moments balance (Andrews, 2017)). With this (non-SE) insight, one then looks for recognition of a "fleet design objective" to address "availability" (given the Swedish Navy's need for several boats to accomplish their specific operational role in the Baltic Sea). As with the previous Australian SE approach, availability/sustainability of a submarine fleet in Coles (2012) terms, could have been incorporated in this OA approach but once again seems to have been omitted from a holistic, but single boat, design consideration.

Work on two current US Navy nuclear submarine programmes was presented to the 2017 RINA conference from NAVSEA's controlling approach. The first by Daley et al. (2017) looked at ESD uncertainty and reducing risk, for the OHIO Replacement (SSBN) Class, while the second by Parker et al. (2017) considered the application of Set-Based Design when applied to ESD of the new submarine attack class (SSNX). The OR Class has already been discussed from a sustainability stance in the last section, but the Daley et al. paper is a more direct naval architecture focused exposition dealing in some detail with the choice of hull dimensions and the issue of weight/stability margins. The OR SSBN paper's main issues are preceded by a brief coverage of "Arrangements" that specifically mentions "sustainability requirements" but without referencing Baker et al. (2015) paper discussed in Section 3 above. Thus "space reservations" are said to be provided to "accommodate quantified variations", which presumably covers added space margins for both maintenance and adaptability/Upgrade TL?

Set-Based Design, the focus of Parker et al., is an established design philosophy in recent NAVSEA surface ship programmes (Dorey et al., 2009). As Parker et al. conclude: "The purpose of using a set-based strategy is to make the right decisions first time…". This seems to be still very FoC design and build focused, with little apparent read across from the OR SSBN Fleet sustainability perception. So, unless "the cost, schedule and capability" emphasis stated gives prominence in the defining of "capability" to the sort of fleet level view of TL ownership that Coles (2012) clearly identified as necessary for the Collins Class SSKs, then again Availability at a fleet level seems to be under called by Parker et al. It is notable that when Coles addressed this for the Collins SSKs, he drew on his own deep involvement in nuclear submarine Sustainability for the UK SSN/SSBN Fleet (Coles, 2013). At the Appendix is a summary by Coles of his response to reading the first draft of the current paper.

The current author in his Submarine Conference paper (Andrews, 2017), pointed out how different submarine design was from even naval surface ship design, went beyond the direct "naval architecture" of submarine design to address the "fuzzier" aspects (see Andrews, 2022 for an explanation of this term) regarding the acquisition and ownership of such vessels. Much was discussed in that paper's Section 6 addressing acquisition, support and the danger of underestimating the risk/difficulties of "repeat design" from a submarine design perspective. Submarine support was seen as a costly endeavour, if (from the publicly available data on the Collins Class) the inadequate provision of resources to both sustain and retrofit necessary upgrades is not comprehended right from early-stage design (ESD). Yet again this author has argued that ESD of complex vessels is not to be rushed in the desire to get into detailed production design and build, as it is only in ESD that for "complex ships" the design team should and can adequately explore sufficiently comprehensively the full scope of capability, cost and risk, Through Life (Andrews, 2018) – and we should add **to achieve the necessary whole class Availability**.

Having by and large criticised even this last series of publications for their overly narrow single boat design focus, the last paper considered in this section (like Coles' papers already referenced (2012, 2013) takes a very broad overview of the ownership of those most demanding of vessels - nuclear submarines. The recent conference paper by Macdonald & Nicholl (2022) is entitled "Nuclear Submarines – the most complex endeavour in defence" and is directly addressing the Australian intent to acquire a fleet of SSNs, with the active support of the USA and UK. These authors argue from their experiences in senior UK MoD positions and subsequent leading roles in the naval acquisition industry, that the wider aspects (in part covered by the current author's comments on the "Fuzzier" half of naval ship design to an American audience (Andrews, 2022)) are what make the design and acquisition and TL ownership of nuclear submarines such a demanding and unique enterprise. They see the complexity as made up of the joint involvement of players from: industry; finance; military; societal; legal and regulatory; and political, and that "not everyone understands all this". Looking to the nuclear submarine endeavour that the Commonwealth of Australia is commencing to embark upon, they identify three critical success factors:

- The right mindset.
- An enterprise approach.

- Continuous investment, indefinitely.

While not arguing with their strategic analysis, that acquiring such a "capability" requires this national commitment, the author as a submarine designer, teacher and researcher and bearing in mind the insights from the analysis in the current paper that the key capability of Fleet availability needs great prominence, I would have liked a greater emphasis in the Macdonald and Nicholl's paper of the unique demands required of submarine designers designing such system of systems, which have to operate in extreme physical environments. Understanding specifically both the physics of pressure hull collapse and manoeuvring at speed (both of which this author has taught at Master's level from first principles) and then integrating the many sub-systems (from nuclear power plant through combat systems to the many fluid service systems – highlighted even at the Concept Phase by Figure 3) in a densely packed and finely balanced vehicle of several thousands of tonnes, which is also home to a hundred highly trained sailors, can be seen as a pinnacle of engineering design. And, I would argue, this specifically requires producing and sustaining an engineering cadre with submarine naval architectural knowledge and experience, if this "indefinite" endeavour is to be successfully pursued.

5. HOW CAN SUBMARINE DESIGN BETTER ADDRESS AVAILABILITY?

While accepting that achieving an acceptable level of availability for a fleet of submarines is a wider commitment than just "good ship/submarine design", whatever that might be taken to mean (see Andrews, 2022), there are suggestions already made in this paper that early recognition in the design process can greatly assist in enhancing the fleet availability when deriving for a new submarine design the "design style" (see Table 1 taken from Andrews, 2021). Table 1 under Design Style includes Adaptability (see Whybrow, 2014) but neither Availability nor Readiness, because it is still addressing the individual submarine rather than the Class/Fleet as a whole, even though it covers elements of the associated term "Sustainability". However in addition, as is clearly spelt out by Coles (2012), many of the issues to achieve adequate, or preferably "best in class", Sustainability/Availability/Readiness are of a very wide scope regarding the whole submarine enterprise of a nation, if that nation is genuinely operating or seeking to operate a modern class of submarines with sufficient "Fleet Readiness".

Table 1. Listing of Style topics relevant to a submarine design (Andrews, 2021)

Stealth	**Protection**	**Human Factors**	**Sustainability**	**Margins**	**Design Style**
Acoustic signature	Underwater weapon effect	Accommodation & Escape	Mission duration	Space	Robustness
Magnetic	Fire	Access	Watches	Weight	Adaptability
Infra-red	Shock	Maintenance levels	Stores	Vertical & Longitudinal centre of gravity	Modularity
Radar cross section	Damage control	Operation automation	Maintenance cycles	Power	Operational serviceability
Visual	Collision	Ergonomics	Refit philosophy	Services	Producibility
	Above water Weapon effect		Upkeep by exchange	Board Margin (future upgrades)	
	Corrosion				

Assuming the widest cultural and governmental commitments are made (as called for by Macdonald and Nicholl (2022)), then at the design level there are fundamental technical choices to be exercised and importantly these have to be made and held to, right from the earliest steps in the crucial first (Concept) phase of submarine design. Thus, one can have a clear availability objective, such as that for the US Navy OHIO Replacement Class': "same availability from 12 new as the 14 SSBNs being replaced" or a specific Material Ready Days for a given fleet, which is based on (and hopefully with improved availability) a previous class's achieved readiness.

However, there are some obvious measures with which a clear availability for the new class can be converted into specific design objectives and translated via a specific set of style choices (see Table 1) into technical features that are crucial in the initial design explorations of options and trade-offs (Andrews, 2018). Thus, space margins may need to be more generous than in past practice, where often arrangements have been too tight in way of equipment that in maintenance evolutions is critical to have easy access or even to be removed. There is also the problem of new equipment or different layout arrangements where the novelty may prove challenging in the life of the class. Adequate space margins in submarines are always fought over as the desire to keep to the limited pressure hull diameters for general space efficiency, means the only way to "gain" space inside the pressure hull is to make it longer. This then leads to consequential greater submarine weight, powering demand and probably BG and LCB impacts (Andrews, 2017). Further knock-ons, such as increased ballast tank capacity to maintain an agreed reserve of buoyancy, show how easy it is to resort to cramped maintenance spaces and make removal routes less than efficient. A focus on availability requires courage from the design manager to allow the submarine's size to grow and defend the likely hike in initial (build) cost by countering that improvement in fleet/class availability is actually the right measure of the overall project's VFM through life.

It has been projected that innovations, such as modularity (Whybrow, 2014), and novel designs, such as those regularly explored in the annual UCL Submarine Design Course (UCL, 2022), including such innovations as hybrid nuclear plant (Bradbeer, 2015), AUV MotherSubs (Pawling and Andrews, 2011, Purton, 2016) and Rim drive (Collins et al., 2015), need to be pursued if future submarines are to exploit new technologies. The author has argued this requires a DBB like approach to ESD used in all the above UCL studies (Andrews, 2018) so that more than gross weight and space balance is sufficiently explored in the Concept Phase. Given this conference's focus on Availability that exploration, for submarine innovation adoption, must not just be in terms of crude overall sizing and initial cost (UPC) but clearly address crucial architectural implications at the sizing stage, as is outlined in the following two paragraphs and Figure 3.

One aspect that recent research at UCL into submarine design has re-enforced a focus on maintainability, as part of fleet /class availability, is to tackle in ESD the implication of the host of distributed ship service systems (DS3). Thus, the DS3 can then be part of enabling key choices affecting maintainability to be explored well before the final concept design is fixed and taken into what is properly designated Feasibility. This is because the next design phase after the Concept Phase is to check the full technical feasibility to ensure that the agreed vessel's overall size and balance won't have to subsequently change, as this is likely to cause further size growth and/or deplete sensible margin provision.

The UCL research into distributed ship service systems (DS3) in ESD exploits the long standing UCL developments in the architecturally/configurational based approach to "ship synthesis" (i.e., the Design Building Block (DBB) approach (Andrews and Pawling (2003), Andrews (2018)). Mukti's (2022) research into DS3 in Early Stage Submarine Design was precipitated by the realisation by the current author that while DS3 contributions are substantial as a vital part of sizing a submarine, it has been difficult to do other than use simple sizing algorithms, such as the weight of a distributed system increases linearly with the length of the pressure hull or the pressure hull volume (Burcher and Rydill, 1994). This then inhibits introducing any variation in the distributed system's arrangement or reinforces the pressure to stick with technologies from past practice or having to adopt potentially excessive design margins for any novelty. The approach to DS3 that Mukti developed utilises a network flow analysis based on a surface ship application pioneered by Brown's research group in Virginia Tech (Parsons et al., 2020, Brown, 2022) that optimises efficient energy flow in power systems or in Mukti's submarine research each of 14 discrete submarine DS3 relevant mode of energy transfer. The marriage of the submarine internal architecture with the routing of the 14 systems, at a broad concept design level of definition, is shown in Figure 3and has been demonstrated by Mukti (2022). Such an investigation undertaken before proceeding beyond the concept design step, could be used to justify sufficient access for maintenance and for the removal of major items at refit and thereby enhance the confidence that the "design price" of improvements in unit availability could be justified as an investment in fleet/class availability.

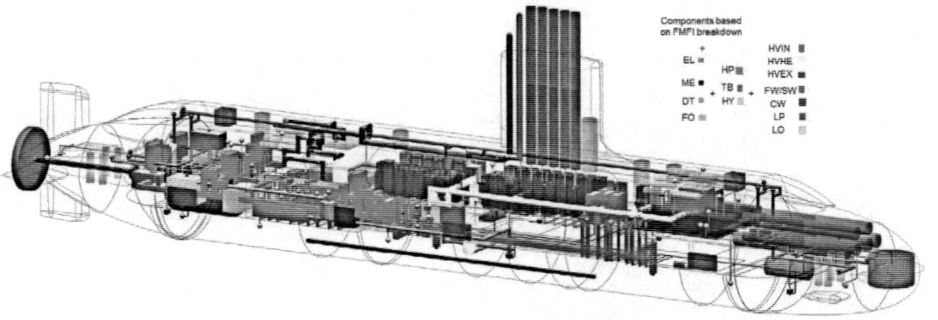

Figure 3: Mukti 's (2022) representation of the Distributed Ship Service Systems for a SSK Concept Design (Refined DS3 model in the third step of the Physical Loop method showing the progress in placing DS3 system by system in the 3D layout. There were 365 Design Building Block objects at component granularity and 472 Design Building Block objects at connection granularity level. The latter shows top level distributed systems routings for 14 services (DS3))

6. CONCLUSIONS

This paper has sought to clarify the conference's two themes of capability and availability, by arguing that availability has been seen, if at all, as part of ILS/Design for Support and, even then, has not been a primary design focus (based on the open literature on warship and, specifically, submarine design). Discussions on submarine design have focused on the main departures from the design of their surface rivals – or "targets' according to the submariners - and how to maximise performance or capability for the available affordability. All too often this has been done in terms of the design of an individual vessel (rather than the capability of a given fleet or class of submarines), whereas availability and readiness only make sense looking at the number of "available hulls" to the Fleet Commander. Or in the unique case of a SSBN force, sustaining the national deterrence capability.

Engineering designers have known for several decades that complex vehicles and systems of systems (submarines being in both categories) should be designed to the principles of Concurrent Engineering. Huthwaite's (1994) seminal text lists 40 "Ilities" or transversal/whole system characteristics sought in the design of complex systems – see also Table 1. Yet he does not include in his list of 40 that of availability, 'though in his defence he does include: Adaptability, Affordability, Deliverability, Compatibility, Installability, Recoverability, Recyclability, Reliability, Removability – all of which have some relevance to availability and more widely to submarine design. (Interestingly, two other clear omission from Huthwaite's "ilities" list are "maintainability" and "sustainability", so does this reveal again the Cinderella status of ILS/DforS in most engineering designer's considerations?)

It could be argued that the sense in the Commonwealth of Australia publishing John Coles' (2012) critique as well as his "get well programme", which highlighted the inadequate availability of their six COLLINS Class SSKs, has made the submarine engineering and design community properly aware of the importance of availability. But more than that, as I hope this paper argues, we have a new submarine design paradigm – we should not be designing a single submarine for its cost-effectiveness (however difficult even that is to encapsulate – see Hockberger, 1993) but rather seek to achieve the best VFM for a new class or even the whole submarine force, where the best sustainability/availability/readiness (for the resources sensibly deployed) is a primary objective for the new design of submarine. One might even take Fuller's "To dive; To move; To live; To survive" (Fuller, 1996) and add: "To be Militarily Available/Ready for the Maximum amount of the Class's Life in Service" – if that can be expressed in the three words "To be available"?

7. ACKNOWLEDGEMENT

This paper was started in response to the topic of Availability posed for Submarines 2023. I shared the draft with my close friend and erstwhile RCNC colleague, John Coles, knowing his leadership in particularly of the Collins Class Sustainability Study that he led in the last decade. John kindly critiqued my first draft and, while the current paper is entirely my authorial responsibility, it has been very significantly enhanced by his comments, many of which I have attempted to incorporate – not least John's emphasis on Sustainability (rather than Availability) and even more so his identification of the need to "maintain the design intent" with the watch word of "readiness' – allowing me to quote from Hamlet!

8. REFERENCES

1. HITCHENS, D K., 'Managing system creation', *IEE Proceedings, Vol 133, Pt A, No 6,* Sept1986.

2. ANDREWS, D., 'Babies, Bathwater and Balance – the Fuzzy half of Ship Design and Recognising its Importance', *SNAME Maritime Convention 2022, Houston, TX*, Sept 2022.

3. ANDREWS, D J., et al., 'SUBCON - A New Approach to Submarine Concept Design', *Warship 96 'Submarines 5', RINA*, June 1996.

4. ANDREWS, D J., 'Submarine Design Is Not Ship Design', *RINA Warship Conference: Naval Submarines & UUV's*, 16. Bath, UK. 2017.

5. ANDREWS, D J., 'Who Says There Are No Real Choices in Submarine Design?', *On-line Conference: Submarines, RINA*, June 2021.

6. COLES, J., 'Study into the Business of Sustaining Australia's Strategic COLLINS Class Submarine Capability', *Commonwealth of Australia, Russell Offices, R1-G-C052, Canberra,* Nov 2012.

8. BURCHER, R. and RYDILL, L J., 'Concepts in Submarine Design', *Cambridge UP,* 1994.

9. UCL, 'Submarine Design & Acquisition Course', *Mechanical Engineering,* Mar 2021. www.ucl.ac.uk/mechanical-engineering/study/other-programmes/submarine-design-acquistion-course

10. DANIEL, R J., 'Considerations Influencing Submarine Design', *RINA Symposium on Naval Submarines,* London, May 1983.

11. ARENTZEN, E S, and MANDEL, P., 'Naval Architectural Aspects of Submarine Design', *TransSNAME, Vol. 68*, 1960.

12. BROWN, D K., and ANDREWS. D J., 'The Design of Cheap Warships', *Proc. of International Naval Technology Expo 80*. Rotterdam, Netherlands, 1980.

12. WROBEL, P G., 'Design of the Type 2400 Patrol Class Submarine', *Trans RINA, Vol.127,*1985.

13. PRINS, C A., and EVERARD, B., 'Approaches to submarine design in a changing environment', *INEC 96,* April 1996.

14. MAIER, M W., 'Architecting Principles for Systems-of-Systems', *Systems Engineering, Vol.1 (4),* 1998.

15. ANDREWS, D J., 'The Sophistication of Early Stage Design for Complex Vessels', *TransRINA Vol.160, IJME Special Edition*, Oct 2018.

16. BETTS, C V., 'Nuclear submarines in the Royal Navy – the design challenge', *BNES Annual Lecture, Nuclear Energy, Vol.38, No.2,* Apr1996.

17. www.paramarine.qinetiq.com Accessed 7/8/2019.

18. ANDREWS, D J., and PAWLING, R J., 'SURFCON A 21st Century Ship Design Tool', *8th International Marine Design Conference*. Athens, Greece, 2003.

19. FULLER, G H., 'To Dive: To Move: To Live; To Survive: The Essentials of the Submarine', *RINA Naval Submarines 5, London,* June 1996.

20. TASKER, P H., and WILCOX, N., 'The Royal Navy ASTUTE Class SSN', *RINA Naval Submarines 6', London,* June 1999.

21. SULLIVAN, P E., and TIBBITTS, B F., 'Naval Submarines', Chapter 56 in LAMB, T. 'Ship Design and Construction', *SNAME, Jersey City, NJ*, 2004

22. SMITH, K R O., and STEINHOLM, B L., 'Platform System Management, Are we at the Limit?', *RINA Naval Submarines 6, London,* June 1996. *RINA Naval Submarines and UUVs', Bath,* June 2011.

23. NICOD, M., 'Impacts of the maintenance on a Submarine Basic Design', *RINA Naval Submarines and UUVs', Bath,* June 2011.

24. SMITH, S., 'Incorporating Through Life Support requirements into Submarine Design', *RINA Naval Submarines and UUVs', Bath,* June 2011.

25. BAKER, W., et al., 'Design for Sustainment: The OHIO Replacement Submarine', *US Naval Engineers Journal, Vol.127, No.3,* Sept 2015.

26. MACDONALD, M., and NICHOLL, A., 'Nuclear Submarines – the most complex endeavour in defence', *SIA Biennial Conference, Australia,* Nov 2022.

27. RYDILL, L J., 'No, the future of the Corps is not in Systems Engineering', *RCNC Journal,* MoD, Bath, 1966.

28. GRIETHUYSEN, W J., Van, 'Marine Design - Can Systems Engineering Cope?', *IMDC*, Kyongju, Korea, June 2000.

29. ANDREWS, D J., 'Marine Requirements Elucidation and the Nature of Preliminary Ship Design', *TransRINA/IJME Vol.153 Part A1,* 2011.

30. MACDONALD, G., 'Submarines, Naval Architects & Systems Engineering', *RINA Naval Submarines 9, Glasgow,* June 2008.

31. ELLIOTT, C., and DEASLEY, P., 'Creating Systems That Work: Principles of Engineering Systems for the 21st Century', *Royal Academy of Engineering, London*, 2007.

32. ANDREWS, D J., 'Ship Project Managers Need to be Systems Architects not Systems Engineers', *RINA Conference on Maritime Project Management, London,* Feb 2016.

33. NORDIN, M., 'A Novel Submarine Design Method', *PhD Thesis, Chalmers Univ., Gothenburg*, 2014.

34. DALEY, S F., et al., 'Managing Early-Stage Design Uncertainty and Establishing policy for Low Risk Design', *RINA Naval Submarines and UUVs', Bath,* June 2017.

35. PARKER, M C., et al., 'Set-Based Requirements, Technology and Design Development for SSNX', *RINA Naval Submarines and UUVs', Bath,* June 2017.

36. DOREY, N., et al., 'What is Set-Based Design', *US Naval Engineers Journal, Vol.121, No.4,* 2009.

37. COLES, J., 'Smart Submarine Sustainment A Perspective', *Submarine Institute of Australia Science, Technology & Engineering Conference, Adelaide, S. Australia*, 2013.

38. WHYBROW, N R., 'Adaptability in principle, in design and in practice – Applicability of this characteristic to the modern Naval Submarine', *RINA Naval Submarines and UUVs', Bath,* June 2014.

39. BRADBEER, N., 'Design of Submarines for Hybrid Nuclear/Battery Propulsion', *Proc 12th IMDC, Tokyo Univ.,* May 2015.

40. PAWLING, R., and ANDREWS, D J., 'A Submarine Concept design – The Submarine as an UXV Mothership', *RINA Submarines and UUVs*, Bath, June 2011.

41. PURTON, I M., 'Concept Exploration for a Novel Submarine Concept Using Innovative Computer-Based Research Approaches and Tools', *Ph.D. Thesis, University College London*, 2016.

42. COLLINS, L. et al., 'A new approach for the Incorporation of Radical Technologies: Rim Drive for Large Submarines', *Proc 12th IMDC, Tokyo Univ.,* May 2015.

43. MUKTI, M H., 'A Network-Based Design Synthesis of Distributed Ship Services Systems for Non-Nuclear Powered Submarines in Early Stage Design', *Ph.D. Thesis, University College London*, 2022.

44. PARSONS, M A., et al., 'Early-Stage Naval Ship Distributed Systems Design using Architectural Flow Optimization' *JSP&D, Vol.37,* 2020

45. BROWN, A J., 'Design of Marine Engineering Systems in Ship Concept Design', *Chapter 1 'Marine Engineering, SNAME, Alexandria, VA.* 2022.

46. HURTHWAITE, B., 'Strategic Design: A Guide to Managing Concurrent Engineering', *The Institute for Competitive Design', Rochester, MI,* 1994.

47. HOCKBERGER, W A., 'Cost and Operational Effectiveness Analysis (COEA) in Naval Ship Design', *SNAME Chesapeake Section,* May 1993.

APPENDIX - COMMENTS ON AVAILABILITY BY JOHN COLES

I have commented on the paper and consider it is over focused on design. Availability or readiness is quite a different capability - it is the sum of wide range of outputs or enablers undertaken in the procurement process - not necessarily directly linked to the development of the design itself. It is much more dependent on key enablers - reliability of machinery is key, so making the correct equipment selection is the issue not specific design issue. Often the design solution (to improve availability) has been to change the machinery - and that is not directly a whole boat issue unless the space demand is significantly reduced - and that would be part of the design process. But you can improve availability in service without undermining the initial design solution. More broadly I suggest that other factors are far more important - these relate to infrastructure, spares availability, a knowledgeable naval base/industrial structure and organisation, etc. All these are not the purview of the naval architectural features of a design but the managerial influences outside.

You will know of course that my work on Collins did not come to any view about design issues affecting the availability, although there were some. The design issue of importance to come out of my review was the absence of a proper system for maintaining the design intent. The greatest improvement in the Collins Class performance was to reduce the time spent in maintenance in major overhauls (by 33%) reduce overruns in planned maintenance (better management of material supply) and improved reliability of some key equipments - none of these issues are related to the design itself - which is under the right leadership and is extremely capable.

Perhaps worth saying surface ship both commercial and naval often have as much as 340 days per year availability - especially cruise ships - and this must be embedded in the design and support infrastructure. Two naval classes had contractual availabilities of over 320 days per year will payment depending on performance. Theses naval ships were at the simpler end of naval vessels and to achieve this performance the crew was at least (from memory) 1.5 times the complement. I doubt if submarines could get as high as this but SSBNs do have a high availability - although I doubt if anyone would release such figures. I do not know what is being achieved today for the RN SSNs because that is not in the public domain.

Finally, the other factor I brought out was the need for all the key agencies helping to deliver availability work as a 'collective enterprise' - In Australia this was achieved because all the key elements were funded directly by the Australian Government - it was the collective leadership that help to deliver my recommendations.
From my point of view, you over focus on design in delivering availability whereas I believe it is more influenced by other external and none design issues. You need a very compelling design change to radically alter the availability.

I hope this helpful. The issue to get across is that availability or readiness need to be inputted to design process with much greater focus - and we can both agree on that!

9. AUTHOR'S BIOGRAPHY

David Andrews FREng, PhD, FRINA, FSNAME, RCNC was given a new Chair in Engineering Design at UCL in 2000, following his early retirement, when a Director (1*), from the UK Ministry of Defence. His early career in the RCNC included several postings in both in-service conventional and new construction nuclear submarine design sections and as Section Head for the VANGUARD Class Structures and Hydrodynamics Design Section. He was subsequently Head of Preliminary Design responsible for all initial naval design studies, including for the Future Attack Submarine. From 1993 to 1998 he was appointed the MoD Professor of Naval Architecture at UCL, where he directed the MSc in Naval Architecture and the post-MSc Submarine Design Course. In 2001 he set up a new Design Research Centre in the Department of Mechanical Engineering, which is focused on CASD and the design of physically large and complex systems. He is a Fellow of RINA, a former Vice President and previous Editor-in-Chief of the IJME. In 2000 he was elected a Fellow of the Royal Academy of Engineering and was until 2022, the International Chair of the triannual International Marine Design Conferences. In 2020 he was awarded RINA's highest award, the William Froude Medal, "for his contribution to the field of ship design" and in 2021 awarded the SNAME David W.Taylor Medal for Lifetime Achievement.

WAKES OF SURFACE-PIERCING CYLINDERS

D Potts and **Z Leong**, Australian Maritime College, Australia, **J Binns** and **A Skvortsov,** Defence Science and Technology Group, Australia, **H Marcollo**, AMOG Consulting, Australia

SUMMARY

This paper discusses the wake patterns of surface-piercing cylinders, which are a relevant model to masts on submersible platforms. The wake generated by a mast is difficult to predict due to the complex interplay of many factors such as mast shape, size, submerged length, velocity, and fluid properties. This paper presents a novel experimental programme utilising the Towing Tank at the Australian Maritime College. The experimental setup involves mounting a cylinder on a submerged platform from which the cylinder would penetrate up through the water column ('bottom-up'). The novel experiments result in a valuable data set for the wakes generated by surface-piercing cylinders which may be used to validate numerical and analytical models. The platform itself provides a unique capability for future experiments, which may include many cylinders with different cross sections.

NOMENCLATURE

AMC	Australian Maritime College
AR	Aspect Ratio
D	Diameter
D_1	Bow wave height
D_2	Bow wave spray peak height
FIV	Flow Induced Vibration
Fr	Froude number
HMAS	His Majesty's Australian Ship
ITTC	International Towing Tank Conference
L	Submerged Length
L_0	Ventilation
L_2	Bow wave spray peak length
L_4	Bow wave spray length
Re	Reynolds number

1. INTRODUCTION

The visual signature of a submerged platform mast primarily comprises of its profile, its wake pattern, and the presence of bubbles (white water) in the wake (see Figure 1). While the profile of masts can be made less visible to sensors by using coatings, the wake generated by a mast is much more difficult to mask, let alone predict due its complex dependency on the mast shape, mast size, submerged length, velocity, and fluid properties.

Figure 1: Conventional submarine at snorting depth, HMAS Farncomb at speed showing the generated wake downstream [2].

The wakes of surface-piercing cylinders were first studied in the late 1940s, where a vast experimental programme was conducted by Arthur Hay [2]. The seminal report consisted of a plethora of data on the force and images of the wake for both above and below the water surface of cylinders of different diameters and velocities. There have been further experimental studies since using better visualisation techniques and hardware on various components of the wake [2]-[7]. These studies however were of a lesser Froude number (*Fr*) compared to Hay and do not cover the full range of expected operational velocities for masts. This is due to two primary factors: the first being the testing facility – the speed of the flow required to increase the Froude number is too great; and the second being the model – typically only a single large diameter is tested to reduce vortex induce vibrations. More importantly, the experimental studies by [2]-[7] have the cylinder mounted 'top-down' – extended from the towing carriage down into the water. This experimental mounting configuration does not replicate the in-situ configuration of a mast mounted within the submarine sail which is 'bottom-up'. Therefore, there is a significant discrepancy in bottom effects between the two configurations and also the inability of the 'top-down' configuration to investigate more representative mast shapes due to structural limitations as masts tend to have a reduced cross-sectional area for its top section.

This paper presents a portion of a novel experimental programme undertaken utilizing the Towing Tank at the Australian Maritime College (AMC) whereby the cylinder was mounted on a submerged platform from which the cylinder models would penetrate up through the water column. This submerged platform is the first and only of its kind, making the data acquired unique to the problem at hand, and furthers the capability of future experiments regarding surface-piercing cylinders. This paper presents a portion of the wake results from the experiments undertaken utilising the submerged platform.

2. WAKE DEFINITION

Wakes for vertical surface-piercing cylinders fall into two regions: above and below the free surface. The existence of the two regions is attributed to the free surface and the multi-fluid interface. The magnitude and scale of the wakes are dictated by the geometric properties of diameter (or cross-sectional area) and length (both submerged and in-air), as well as the velocity. The following lists and briefly describes the distinct characteristics within the two regions, which are then visually represented in Figure 2.

- Bow Wave (D_1) – the run-up of fluid of the fore area of the cylinder in the direction of travel. Height is the primary characteristic of the bow wave.
- Bow Wave Spray (D_2, L_2, L_4) – a derivative of the bow wave where the fluid (water) separates from the cylinder. The maximum height, the downstream location of the maximum height, and the overall length are the characteristics of the bow wave spray. There is a transition point with respect to the velocity at which the bow wave spray is no longer a derivative of the bow wave.
- Ventilation (L_0) – a pocket of air which forms below the free-surface on the leeward side of the cylinder. The depth of this pocket is the measured characteristic of ventilation. In the instance of the air pocket reaching the submerged tip or end the body is considered to be 'fully ventilated'.

The length of the cylinder is critical to the development and formation of the wakes. The submerged length (L) is presented as a ratio to the diameter and is more commonly referred to as the aspect ratio (AR), where the instance of full ventilation occurs when $L_0 = L$. The in-air length of the cylinder is of secondary importance as it merely needs to exceed the predicted bow wave height.

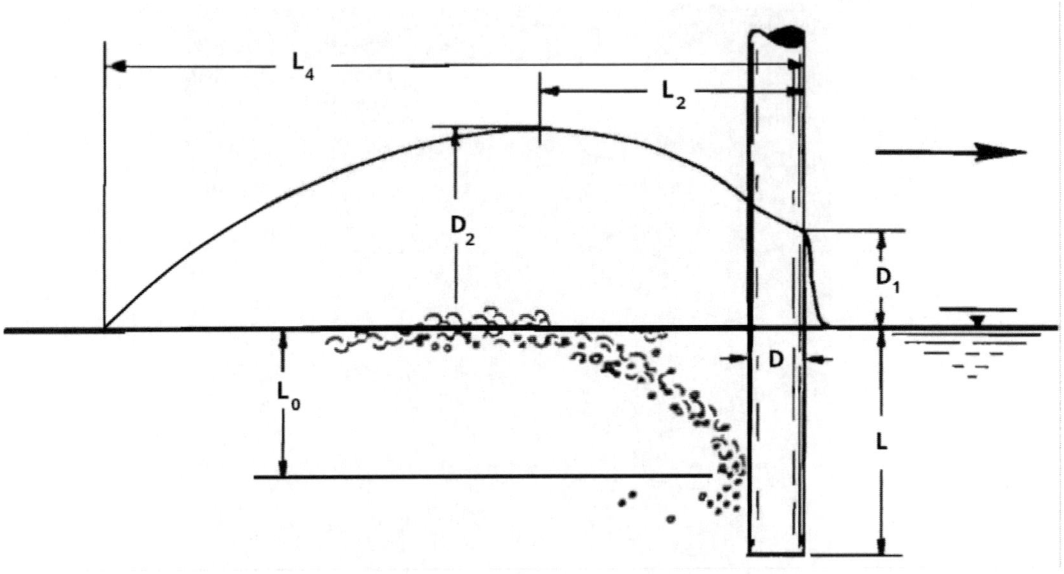

Figure 2: Wake parameter definition visualisation, after [2]

3. EXPERIMENTAL METHODOLOGY

The historical experimental method for vertical surface-piercing cylinders has been exclusively conducted by mounting the cylinder to some form of measuring apparatus above the free surface. However, this method does not readily replicate the real-world scenario of a cylinder protruding up through the free surface from a submerged platform, as such the authors designed and implemented the first ever submerged test platform so as to reflect the real world. With the use of the novel submerged test platform comes the ability to compare and contrast the two methods and determine the differences and similarities in various aspects such as forces, oscillations, and wakes. At the time of writing, the comparison is still ongoing. This paper focuses on the methodology and results of the submerged test platform.

3.1 FACILITY

The facility used to conduct the experiments was the Towing Tank at the Australian Maritime College. The tank has the dimensions of 100L x 3.55W x 1.5D m and is filled with fresh water with a temperature of 18°C ± 0.5°C. The towing carriage has a maximum speed of 4.6 m/s, and the speed measurements are within the recommended International Towing Tank Conference (ITTC) tolerances.

3.2 EXPERIMENTAL SETUP

Figure 3 presents the novel test platform and the rail mounting system. The platform was attached to the towing carriage via a breakaway link which would trigger at an extreme force, disconnecting the submerged platform from the carriage. Three cylindrical models of varying outer diameters (*D*) were tested: 48.3-, 76.1-, and 101.6 mm. For each diameter a series of aspect ratios were targeted by varying the submerged length, see Table 1. The largest diameter is one of the most common diameters cited throughout literature sources.

The cylinders were rigidly mounted to a force transducer, which was mounted below the front end of the platform, so as to measure the forces acting on the model. An accelerometer was mounted to a plate and attached to the in-air free tip, where the subsequent accelerations could be converted into motions. A series of cameras were mounted both above and below the water surface to capture the various wakes features outlined in Figure 2. The above surface cameras were mounted to the carriage, while the subsurface camera was at a known fixed location down the length of the tank.

Figure 3: Submerged experimental platform.

3.3 TEST MATRIX

A broad test matrix was developed to encompass various combinations of diameter, aspect ratio and velocity. The range was selected for comparative purposes against both the literature and the authors own data set for the traditional experimental setup – the cylinder is mounted from above the free-surface and penetrates down into the water column [9] – for the purpose of quantifying the difference in methods and determining the point at which both methods produce the same or similar results. Table 1 presents the diameter and aspect ratio matrix, where within each of the conditions the velocities were within a range of 1 – 4 m/s.

Table 1: Test matrix

Aspect Ratio		**1**	**2**	**4**	**6**	**8**	**10**	**12**	**14**	**16**	**18**	**20**
Diameter [mm]	**101.6**	X	X	X	X	X	X					
	76.1	X	X	X	X	X	X	X	X	X		
	48.3					X	X	X	X	X	X	X

4. RESULTS

Herein presented are sample stills from the experimental programme undertaken for each of the diameters (Figures 4, 5, 6 and 7) where each of still sets is for the same aspect ratio and velocity. The sample stills for each of the diameters show the shape and magnitude of each of the various wakes, where the bow wave and bow wave spray are visible within the near and near far field, whereas the subsurface ventilation is primarily within the far field region as the cavitation induced bubbles remains prevalent in the downstream region. The in-air length of the cylinder as shown in Figures 5 and 6 far exceed the bow wave height, and effectively represent an infinitely long cylinder, however Figure 7 presents a potential high-speed scenario where the bow wave exceeds the in-air length of the cylinder. The bottom left image for each of the figures depicts the subsurface view of the cylinder and captures the underwater phenomenon of ventilation, cavitation and vortex shedding. The vortex shedding is represented by vertical white lines downstream of the cylinder, which are von Karman streets. For the case of the smallest diameter cylinder (shown in Figure 5) the vortex shedding is more prominent when compared to the larger diameters.

Figure 4: Left: 48 mm diameter cylinder, Froude ≈ 1.45 and Reynolds ≈ 4.59E4.
Right: 101.6 mm diameter cylinder, Froude ≈ 1.00 and Reynolds ≈ 9.64E4

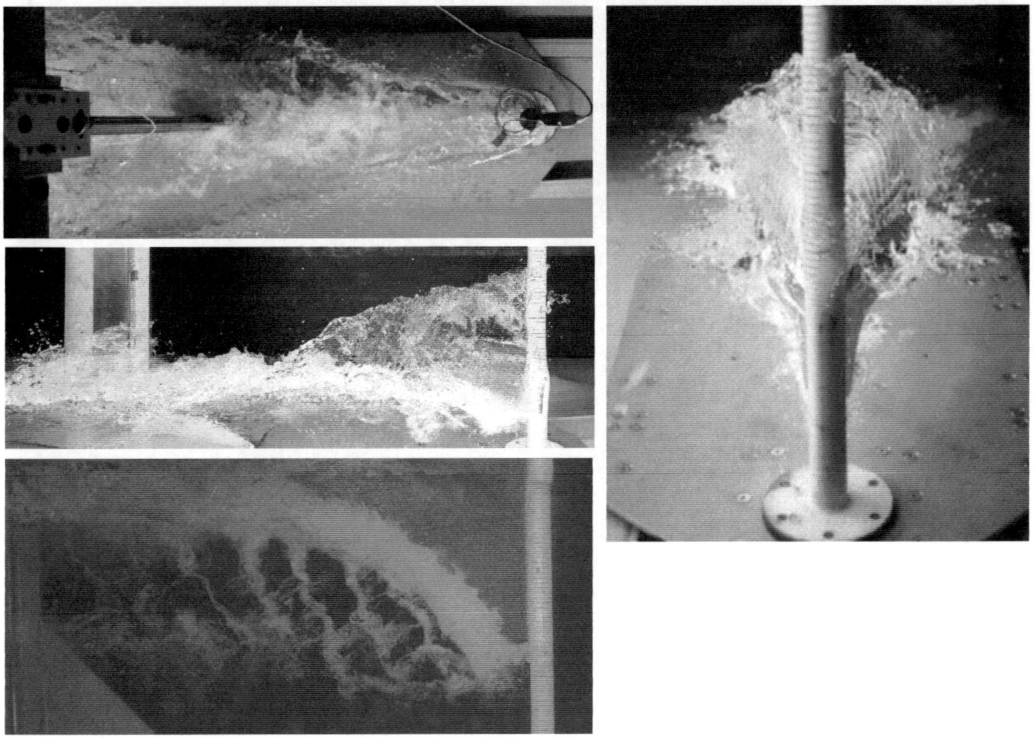

Figure 5: 48 mm diameter cylinder at Froude ≈ 5.8 and Reynolds ≈ 1.83E5

Figure 6: 76.1 mm diameter cylinder at Froude ≈ 4.6 and Reynolds ≈ 2.89E5

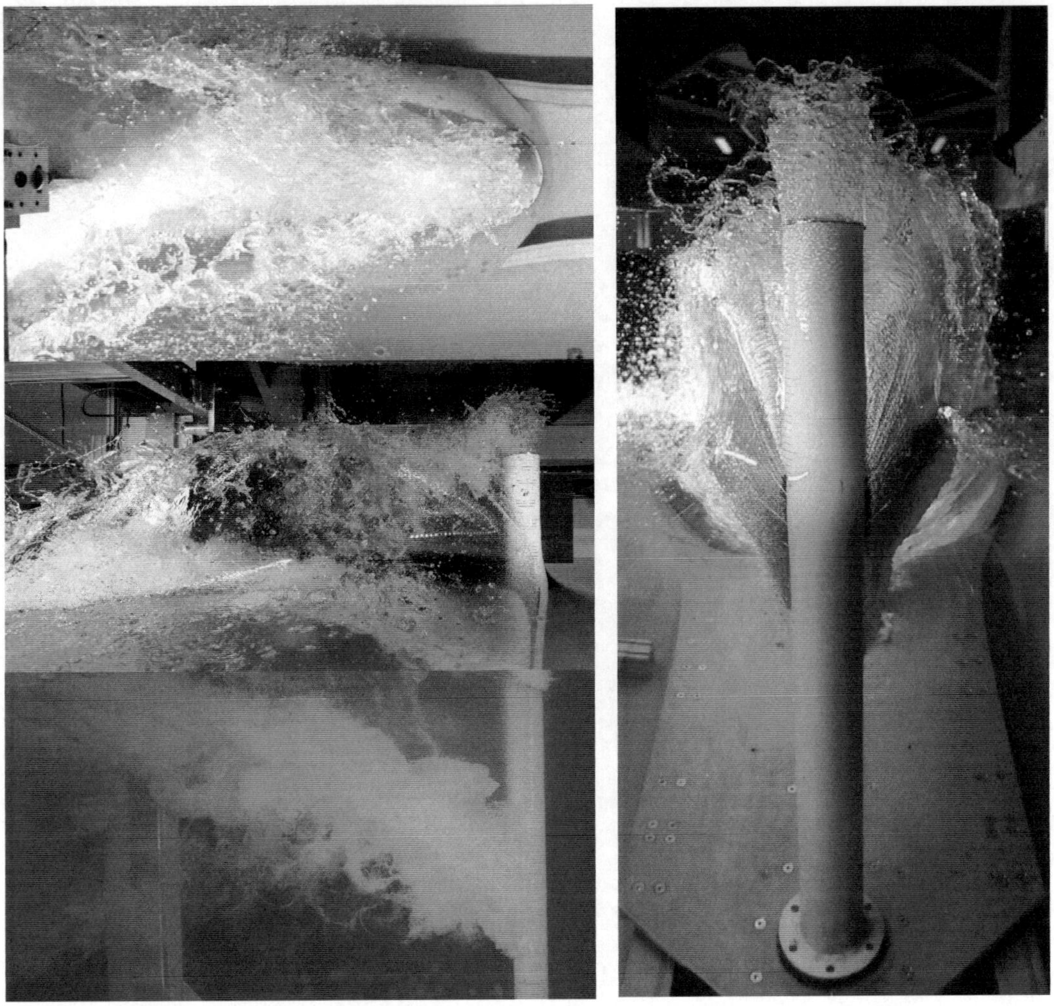

Figure 7: 101.6 mm diameter cylinder at Froude ≈ 4.0 and Reynolds ≈ 3.86E5

Plots of the above surface wake parameters are presented (Figures 8, 9, and 10) where the wakes have been nondimensionalised with respect the appropriate length parameter, which for the above surface wakes is the diameter, and the below surface wake is the submerged length. Each plot contains the three diameters at the same aspect ratio across the velocity range. The black line on each of the plots is the Bernoulli equation for inviscid flow, which has been established as an effective limiting line [8], [9], whereby real-world and experimental results should fall below the line due to energy dissipation caused by friction and fluid resistance. Identification and quantification of the energy dissipation is a new concept which is empirically derived for a single diameter cylinder [8]. The trends for all the plots show that there is a somewhat linear relationship between the nondimensional wake and the squared Froude number. The linear slope of the results appears to change somewhere between Fr_D = 10-15, where the slope changes, as it is within this range that the viscous energy dissipation becomes significant.

Figure 9 and 10, plot zero values at low Froude numbers, this is due to the flow around the cylinders being pressure related where there is no visible flow separation as shown in the top images in Figure 8. Measurement of the ventilation is difficult at lower velocities, hence the zero values shown in Figure 11, where similar to the above surface wakes the flow is pressure related.

Figure 11 presents the nondimensional ventilation for the same conditions as noted above, where again the Bernoulli limiting line has been added. It is important to note that there is an additional fixed limitation (as noted earlier), which occurs when L_0/L = 1, as is the case when the air pocket extends down the entire submerged length ($L_0 = L$). From the results presented, the full ventilation phenomenon does not occur, whereby the submerged portion of the models may be deemed infinitely long. However, it is more than likely that at higher Froude numbers that the air pocket reaches the base and aspect ratio presented is no longer deemed infinitely long.

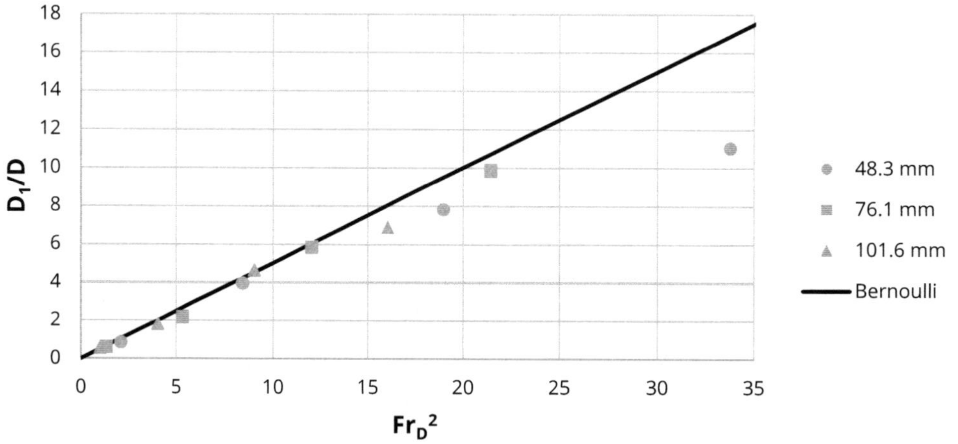

Figure 8: Nondimensional bow wave height with respect to Froude diameter squared.

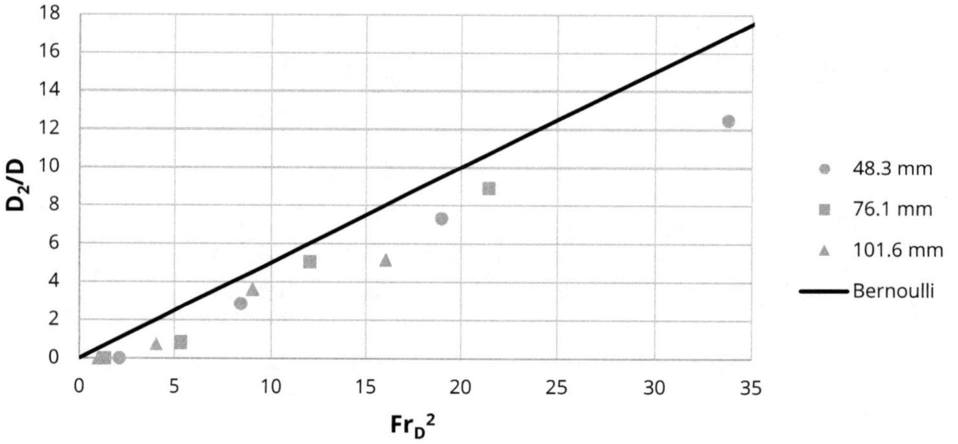

Figure 9: Nondimensional bow wave peak height with respect to Froude diameter squared.

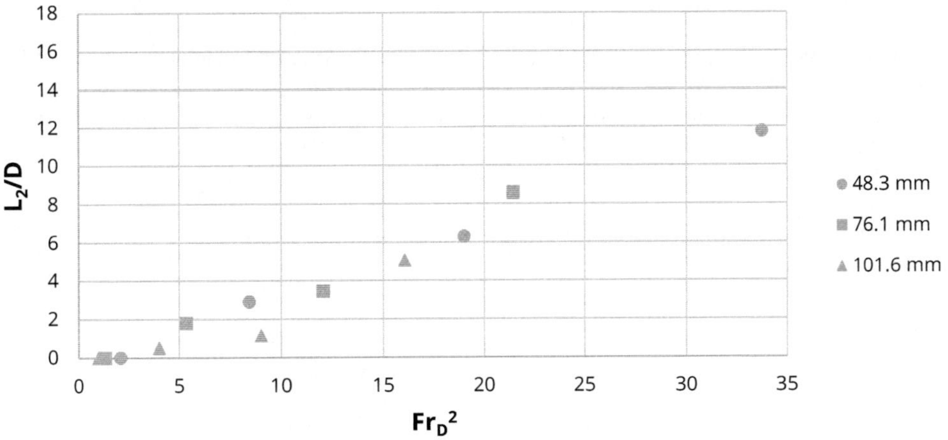

Figure 10: Nondimensional bow wave peak length with respect to Froude diameter squared.

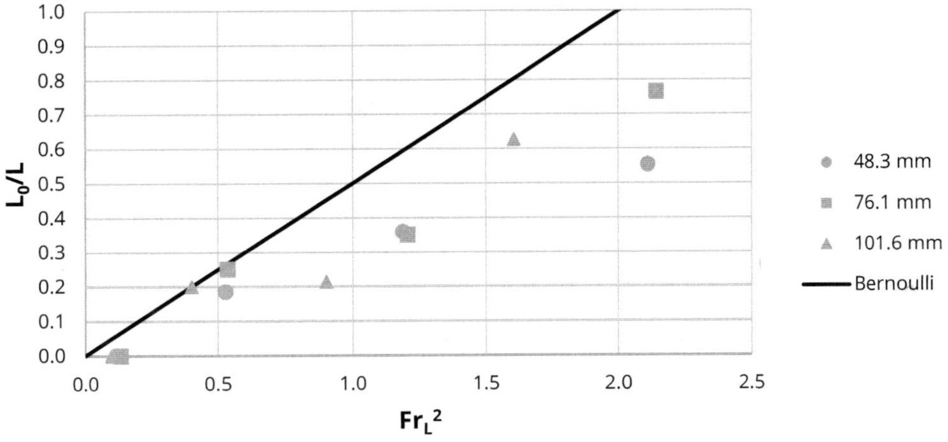

Figure 11: Nondimensional ventilation pocket depth with respect to Froude length squared.

5. CONCLUSIONS

The results from an experimental programme utilising a novel submerged platform – which is representative of the real-world submerged platform scenario – provides a baseline data set which can be implemented in the validation of numerical models, in particular the near field. The use of uniform bare circular cylinders provides the baseline to compare against alternate omnidirectional geometries, such as helical strakes or shrouds, as well as other geometrical changes including a diameter change.

The wake characteristics both above and below the free surface tend to be somewhat proportional to the squared velocity – as in conjunction with the Bernoulli equation – where there is an energy dissipation for the above surface wakes, for which is yet to be fully accounted. Empirical derivations of this relationship may be determined from the experiments undertaken by the authors, which may then be used in validating numerical or analytical models.

The subsurface ventilation has significant implications in the residual wake in the far field downstream due to the ventilated pockets, and as such requires further investigation where alternate geometries may require more extreme conditions for the inception of ventilation and the potential for vortex core cavitation.

6. ACKNOWLEDGEMENTS

The authors would like to acknowledge the following organisations and personnel:

- The AMC Towing Tank Crew, in particular Kye Curwen for jumping into the water as required.
- DSTG for the financial and intellectual support through DSPRA9454 and DSPRA9455.
- A/Prof Gregor MacFarlane, Dr Nick Johnson, and Shannon Keough for their advice and aid both prior and throughout the conduct of the experiments.

7. REFERENCES

1. CONWAY, A S T., 'Investigation into Wakes Generated by Surface Piercing Periscopes', *Doctoral Thesis, University of Tasmania, Australian Maritime College, Launceston, TAS, Australia*, 2017.

2. HAY, A D., 'Flow about Semi-Submerged Cylinders of Finite Length', *Princeton University, Princeton, NJ, USA, Technical Report NObs-34006*, Oct. 1947.

3. AGEORGES, V., PEIXINHO, J., and PERRET, G., 'Flow and Air-Entrainment Around Partially Submerged Vertical Cylinders', *Physical Review Fluids*, vol. 4, no. 6, Jun. 2019, doi: 10.1103/PhysRevFluids.4.064801.

4. ARAI, J., and MORI, T., 'Observations and Particle-based Simulation of Air-entrainment around a Surface Piercing Cylinder', in *MARINE 2021*, Jun. 2021, p. 10.

5. ELHANAFI, A., DUFFY, J., MACFARLANE, G., BINNS, J., and KEOUGH, S J., 'Inline Forces and Bow Wave Height on a Vertical Cylinder Moving in Waves — Experimental Study and CFD Validation', *Journal of Fluids and Structures*, vol. 107, p. 103387, Nov. 2021, doi: 10.1016/j.jfluidstructs.2021.103387.

6. HILO, A K., HONG, J W., KIM, K S., AHN, B K., LIM, G S., And SHIN, S., 'Experimental Study on the Flow Patterns and Acoustic Characteristics of Surface Piercing Cylinders', *Ocean Engineering*, vol. 259, p. 111895, Sep. 2022, doi: 10.1016/j.oceaneng.2022.111895.

7. SAKATA, K., HIRABAYASHI, S., YOSHIMURA, Y., SUZUKI, H., and GONÇALVES, R T., 'Free-Surface Effects on Hydrodynamic Forces of Half-Submerged Circular Cylinders with Low Aspect Ratio', in *COB 2019*, Uberlândia, MG, Brazil, Oct. 2019, p. 8.

8. KEOUGH, S J., *ET AL.*, 'Time Resolved Measurements of Wake Characteristics from Vertical Surface-Piercing Circular Cylinders', in *20 AFMC*, Perth, WA, Australia, Dec. 2016, p. 4.

9. POTTS, D A., BINNS, J R., MARCOLLO, H., and POTTS, A E., 'Hydrodynamics of Towed Vertical Surface-Piercing Cylinders', in *OMAE2019*, Glasgow, Scotland: American Society of Mechanical Engineers (ASME), Jun. 2019, p. 15. doi: 10.1115/OMAE2019-95109.

8. AUTHORS' BIOGRAPHY

Douglas Potts is a PhD candidate in the field of maritime and ocean engineering at the Australian Maritime College. The topic for his thesis is with respect to the Wake Topology of Vertical Surface-Piercing Cylinders. Douglas has been in the research engineering field for over 8 years, with a focus on experimentally driven results with real-world applications in the fields of cable mechanics, vortex-induced vibration (VIV), and hydrodynamics of multi-phase cylinders. Douglas has conducted numerous experimental campaigns utilising the AMC towing tank and the Monash FLAIR Lab for both standard and novel experimental methods.

Dr Zhi Leong serves as a Research Fellow at the Australian Maritime College. His research includes varied applications of Computational Fluid Dynamics and Experimental methods to investigate hydrodynamic challenges in Maritime Engineering, with a special interest in underwater vehicles.

Dr Jonathan Binns is currently a group leader at DSTG for Naval Architecture and Platform Systems Analysis. Jonathan has a PhD in hydrodynamics. Jon held various senior academic and professional roles at AMC, where he led a range of maritime engineering projects. He has specialised in hydrodynamic analysis of performance of maritime platforms.

Dr Alex Skvortsov is currently a Principal Scientist at DSTG for Acoustic Signature Control. Alex has a PhD in sound-vortex interaction and flow acoustics.

Dr Hayden Marcollo is a Director of AMOG Technologies (AMOG's R&D arm) and AMOG Consulting Ltd. Hayden has a PhD in offshore engineering, and held a postdoctoral research position at MIT, which included both physical and numerical modelling of offshore riser systems.

Warship Conference
31st May – 1st June 2023
Guildhall, Bath, United Kingdom

CAPABILITY, AVAILABILITY, COST; THE TRADE AND THE BALANCE

G Wainwright, Babcock International, UK and **R Chaplin,** BAE Systems Submarines, UK

SUMMARY

This paper investigates how capability, availability and cost form the trade basis of a balance that must be struck to deliver a submarine design with agility and pace. It explores how only through continual and close collaboration between the key Industrial Participants can availability be assured with the confidence required to underpin any capability essential for a modern submarine deployed fighting force.

This paper identifies several elements, including the embodiment of 'Supportable Design Principles', that need to be progressed through collaboration between Industrial Participants to assure availability. This paper also looks into future considerations, more specifically around data and digital opportunities for assuring availability through life. It identifies a number of constraints and limitations on these opportunities relating to asset compatibilities that will need to be managed.

1. INTRODUCTION

Submarines are a critical element in any modern, multi-dimensional naval force, capable of conducting covert operations as part of a broader Strike Group or to operate independently and in complete isolation to deliver a variety of warfighting missions. The UK Ministry of Defence (MoD) has previously considered submarines to be the most complex machines ever created [1], however this comes at a price, both in terms of monetary cost and specialist skills to operate and maintain.

This paper explores the availability, capability and cost trade space within design. It discusses how the evidence provision into any decision making process is essential to ensure balanced decisions are made as well as how the maintenance and operating cycle and Support principles form the basis of the Support requirements and rules to influence design. It looks into the organisation structure of integrated teams and how efficiencies and improvements can be realised through proper integration of teams, process and reporting metrics across the whole lifecycle of a submarine. This paper concludes with a review of key developing technology in the digital domain, specifically condition monitoring and additive layer manufacturing capabilities, which can enable optimisation of availability and cost.

2. SETTING THE SCENE

Information relating to military submarines is classified to ensure a sustained operational advantage over a potential adversary. The UK currently has 4 Vanguard Class nuclear powered ballistic missile submarines (SSBN as per the North Atlantic Treaty Organisation (NATO) designation), that deliver the UK's nuclear deterrent. Operation Relentless has ensured that there is always an SSBN on a Continuous At Sea Deterrent patrol. From this it can be deduced that the absolute minimum Operational Availability A_O required to sustain this tempo is 25%. To allow for periods when a submarine is in major upkeep and out of the operational cycle, or in the event of a strategic shock that unexpectedly reduces the fleet size, it may be preferable to target a minimum A_O of 33%. Consequently, for the purpose of this paper a baseline A_O of 40% shall be taken.

It can therefore be deduced that a submarine with a 40% A_O will have a 60% non-availability. This 60% can be split into 'force generation' and 'maintenance' where 'force generation' are elements of A_O that do not relate to the material state of the submarine; crew training, weapon loading, storing ship, etc. and 'maintenance' are elements of A_O that do relate to the material state of the submarine; planned maintenance, defect rectification, capability enhancement or addition, etc. In UK parlance the time taken to restore the material state of the submarine through the 'maintenance' elements are termed the material element of readiness, or 'R' Level [2] with R0 being at Immediate Readiness and ready for task.

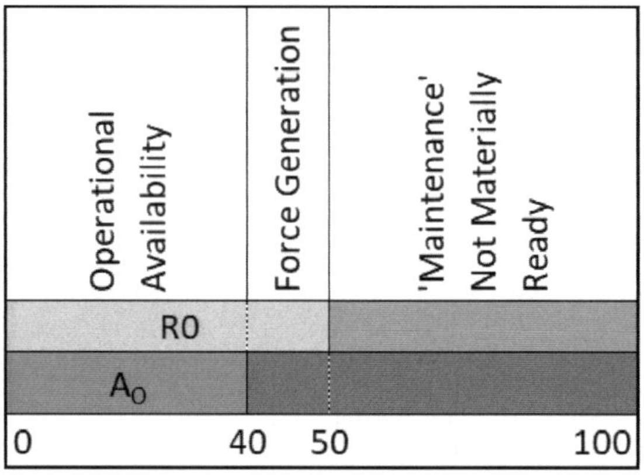

Figure 1: Apportionment of A_O and Readiness Level

From inspection it can be deduced that the vast majority of this 60% non-availability relates to the material state of the submarine. This view is shared by the late Dr Eric Grove who stated that the top two challenges facing the Royal Navy Submarine Service in recent times are "maintaining the operational availability of what we have and maintaining enough personnel to [crew] them" [3]. Given the complex nature of submarines, maintainability should be considered as critical as capability when it comes to overall asset performance; a submarine with one element and not the other is just a target.

3. THE TRADE SPACE

The trade space for submarines has 3 distinct elements: capability, availability and cost. The prioritisation of any element usually results in a sub-optimisation of the other two. A submarine with a world class capability designed with large maintenance envelopes and clearly defined and un-obstructed removal routes will likely be larger in size than its type contemporaries and thus will be more expensive to procure. The infrastructure required to maintain such a large submarine will also be more costly. A submarine of a smaller size that is more affordable must balance capability and availability. The addition of specific capabilities will take up valuable real estate on-board and erode the required maintenance envelopes thus making the submarine more difficult to maintain.

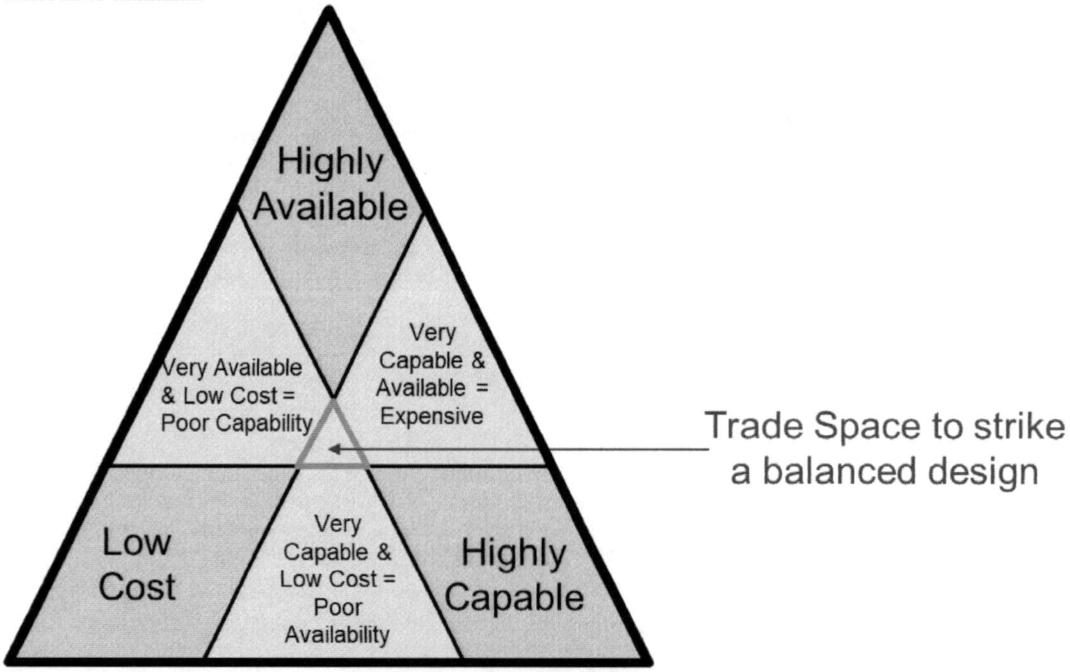

Figure 2: Availability, Capability and Cost Trade Space

3.1 CAPABILITY

In defence the required capability is the critical focus of senior military officials and politicians; "what capability does the platform have and will it have an operational advantage over any potential aggressor". This 'operational advantage' is expressed through capability as a function of asset performance versus hostile threat. The performance of UK submarines needs to continually improve through developments in science and technology [4] as potentially hostile actors continually improve the performance of their respective assets. Any slowdown in UK performance improvement could result in the loss of that 'operational advantage'. Whilst treated in this paper as a trade-off it should be acknowledged that the Support, both in terms of the inherent Design Supportability characteristics and the Support Solution, is a key contributor to the capability to get to, and stay in, the fight to deliver operational effect for any defence asset.

The planned capabilities of future submarines are withheld from the public domain to sustain that 'operational advantage' by preventing hostile actors from pre-emptively developing countermeasures against these future assets. Using the Astute Class nuclear powered attack submarines (NATO designation SSN) as the baseline, the Royal Navy require SSNs to have a world class suite of sensors, fire heavy weight torpedoes and launch Tomahawk Land Attack Missile(s) [5]. This combination will ensure the Royal Navy retain operational advantage and successfully engage the competition. SSNs are also an essential component of Carrier Strike Groups [6]. The UK has committed to funding for the next generation of SSNs [20] to guarantee our security well into the second half of the century.

3.2 COST

Submarine programmes are incredibly costly and the design, build and in-service support of these complex machines consumes a not insignificant percentage of the defence budget. Establishing costs for the respective design, build and in-service phases of the lifecycle is challenging for various reasons [7]. These include costs relating to embarked weapon systems, costs relating to life extension programmes which were not initially planned for during the procurement phase, investment costs for infrastructure and personnel which are required to sustain the strategic base and should not be attributed to specific submarine Classes. There is also the personnel costs for designers, engineers, project managers, production teams, testing and commissioning teams, disposal teams and the crews themselves. There may also be opportunities to releasing intrinsic value during disposal if scrap materials are of sufficient value to warrant recovery.

Using the Vanguard Class submarines and the Trident D5 Submarine Launched Ballistic Missile as a cost comparator, the total procurement cost in 2021-2022 prices has been put at approximately £20bn [7] however some Trident costs were not included in this estimate. The 2010-2025 Trident operating costs have been estimated to be £38.4bn [7] however there may be some additional costs that are not included in this estimate. As a comparison the in-service costs in this instance are approaching double the procurement costs. Using alternative methods for estimating, other sources estimate the total cost of the UK's nuclear weapons programme between 2019 and 2070 to be £172bn [13]. In the US, one study found that retiring five Los Angeles class SSNs would save $1.6 billion in refuelling costs and $250 million annually in Operation and Support costs [14].

Delays in the delivery of maintenance are not uncommon in Nuclear Submarines around the world [15, 16], and this has a perceived waste associated which, for such expensive defence programmes, can bring political pressures. Besides monetary costs associated with the procurement of equipment, facilities, labour etc., one can also attach a cost associated with when the asset in question is not in-use, i.e. a theoretical cost associated with unplanned unavailability. The budget holder pays for equipment to fulfil a specified purpose, when it is not available to the user the asset is unable to fulfil that purpose, and therefore waste is incurred, exemplifying the term 'Value-in-use".

To explain the cost of this waste we can use the availability performance of a system:

Availability (%) = Actual uptime/ planned uptime, where 'uptime' = R0

One can calculate the 'wasted' costs of a system by multiplying the procurement and running costs over the lifecycle by Availability delta from 100%. i.e. if the budgeted lifecycle costs for an asset was £100M and its maintenance cycle delivered uptime of 10 days, but the actual uptime was 9 days, the following wastage costs are calculated:

(1-(9 days/10 days))*£100M = 10% * £100M = £10M.

In other words, wasted costs are the budgeted lifecycle costs multiplied by the overall unplanned downtime as a percentage as shown in Figure 3:

Figure 3: Example of 'wasted costs' (not real data)

This could be translated in to the real-world example of gym membership. For example, a member pays £60 per month with the intention of going to the gym 10 times per month and after a year they terminate their membership. This can be translated in-to a cost of £720 per year for 120 gym visits, or £6 per visit. Due to other commitments/ injuries the member only manages to go to the gym on 100 occasions throughout the year. This means they have wasted (1-(100 visits/120visits)*£720 = £120.

Learning can be taken from legacy programmes undertaken by all nuclear submarine hosting nations to identify design change which would mitigate predicted availability shortfalls. However, this may come at a cost either in monetary terms or on other attributes the product is measured on. The budget holder needs to define an exchange rate between the key characteristics that are being traded so that 'optimal asset value' can be delivered.

3.3 AVAILABILITY

Owing to the impact on defence capability, perceived costs of maintenance delays described earlier and consequential political pressures, the availability of submarines is a critical performance characteristic through the life of the submarine. After all, there is little point in having a capable design if it is not available to the user to get to the fight and stay in it. It cannot truly be measured until the end of life after the Fleet Non-Operation Date, or at least until a point in the life cycle where sufficient confidence exists in what could be considered to be the norm, for example post mid-life refit for submarines. It can however be predicted with varying degrees of confidence.

During the design phase initial assurance can be attained through introduction of availability derived design requirements, appropriately weighted to ensure they are not overlooked, with increased confidence gained through the supportability analysis undertaken once the design has achieved sufficient maturity. Availability is then realised through the effective implementation of the Support Solution. This includes the execution and completion of all planned maintenance; as well as through an agile Maintenance Service Provider with sufficient data and float in the resource profile to undertake the timely rectification of emergent defects within the planned time period [8].

The Support Solution is very complex in itself with lots of moving parts, many different stakeholders and compliance demands from various quarters. The most cost-effective and efficient mechanism to ensure the emerging availability of submarines meets those expected and required by senior defence officials is to robustly couple the submarine design to the Support Solution.

This in itself is becoming increasingly challenging owing to:

1. Increased capability and therefore complexity in design driving additional maintenance which includes software in order to assist the operator;
2. The requirement to adhere to more stringent standards for platform safety when compared to other naval assets, preventing them from deploying until required preventive maintenance items and inspections have been completed;
3. Higher regulatory standards regarding personnel and site safety during maintenance further complicate the issue.

Whilst it may be logical to split the lifecycle at the in-service point as different key Industrial Participants have different responsibilities and the liability for the submarine has changed from private to public ownership, it allows for the specific lifecycle prioritisations. For example, adding capability into a fixed length submarine without necessarily understanding the impact of the sub-optimisation on the other lifecycle phases. To that end, the platform design and the Support Solution design must be integrated, coordinated and matured in tandem. The different elements of the whole lifecycle need to be brought into a harmonious and efficient relationship.

If we are to be successful in our endeavour, tackling the 'Availability' challenge for Submarines needs to be considered from the Enterprise perspective. In the case of Attack Submarines, any planned maintenance maybe balanced and deconflicted to ensure maintenance of the Continuous At-Sea Deterrent.

4. ACHIEVING THE BALANCE

The most effective way of achieving the balance and striking an efficient relationship between the lifecycle phases is to bring together the various stakeholders that have differing responsibilities across the lifecycle into a single integrated and coherent team. This may be difficult to achieve in the short term due to entrenched cultural differences and perceived threats to each other's business model or poaching of qualified resource employed by each organisation. However, the lifecycle for complex assets such as submarines is a long-lasting endeavour spanning several decades, spending time to establish a collaborative culture with all stakeholders fully bought in to a "best for programme" ethos enables the successful integration of organisations with competing objectives to be achieved.

Through this collaborative approach Support can be baked into contract structures and subsequently into the DNA of the design which will allow for obligated availability metrics to be reported. Furthermore pull through initiatives from previous Classes reduces non-recurring engineering costs but these also carry the risk of pulling through supportability issues which may or may not be known at the time. The following sections discuss key factors to achieving the balance.

4.1 DECISION MAKING PROCESS

Through this integrated team the barrier between the lifecycle phases can be broken down such that the transition to service point will be more drawn out and less transactional or abrupt. Support discriminators can be brought into the early design decisions that are being made such that decision makers can take an informed position for each presented option and draw a balanced view on which option to proceed with. This informed view is dependent on the lived experience from the waterfront, what Support implications there may be with each option and what ongoing mitigations may be required as part of the Support Solution. As well as the Support element, this decision-making process needs to include an assessment of the emerging capability, the comparative costs of each option, an assessment on 'buildability' along with any risks relating to sustaining the strategic supply chain and any obsolescence risks if a business were to fail.

A key part of the 'informed position' for decision makers must be based upon objective and repeatable analysis using available data. Using component failure data from previous applications analytical models can be produced that enable a prediction of system reliability and subsequently, availability once combined with predicted time for preventative and corrective maintenance and, logistic delay time for the system. Integration of the system level analysis enables a prediction for overall platform availability. Such reliability models can be produced for various design options to provide a view of which is optimal from an availability perspective.

4.2 MAINTENANCE AND OPERATING CYCLE

The Maintenance and Operating Cycle (MOC) is the top-level requirement for availability. The submarine is required to be available for tasking as required by this cycle. As such the MOC needs to be decomposed into a series of design requirements placed on the platform. It is essential that the design is continually stressed against the MOC and the constituent scenarios that sit underneath it. This will form the basis for the Supportability Case Report and the level of confidence that can be taken that the platform can be a force projection asset. If this decomposition does not occur and the design is not driven by the MOC then there is a significant risk that equipment and system failures occur mid-mission.

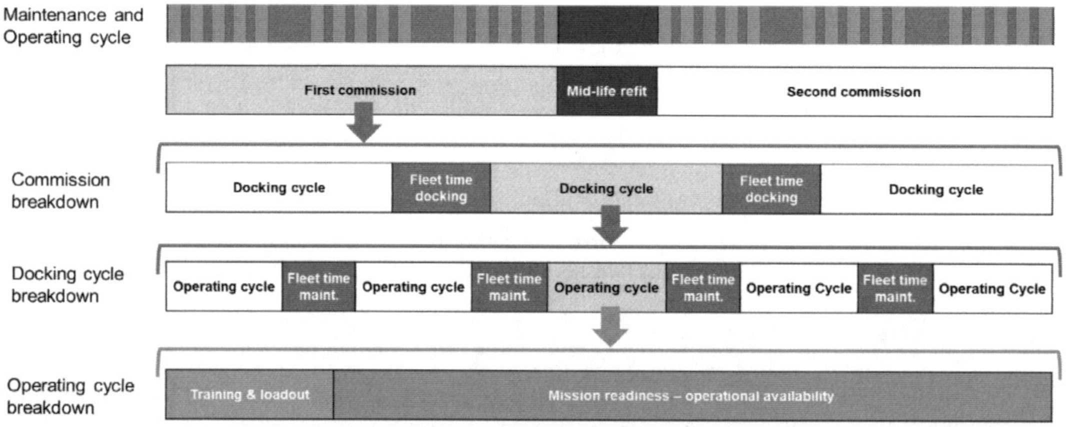

Figure 4: Maintenance and Operating Cycle

Figure 4 provides an overview of a typical MOC for a Submarine. At the most basic level the life of a submarine is two commissions separated by a mid-life refit, which typically includes Class modifications, upgrades, concessions, survey requirements as well as planned maintenance. Traditionally, the mid-life refit involves a major strip of internal panels, i.e. deck heads and linings, non-structural bulkheads and equipment removal as work in way to enable safe access for maintenance, and survey. This may include Hull and Tank plating cuts.

Each commission will likely contain fleet time docking periods, or docking cycles where the submarine will go into the dock for maintenance that cannot be performed while alongside. Periods between dockings will include fleet time maintenance when the platforms are alongside and operational cycles, these operating cycles take into consideration factors that ensure both platform and crew are ready to deliver the required capability, i.e. 'Operational Availability'. The aim is to optimise the periodicity and duration of maintenance in order to maximise Operational Availability.

It should be noted that a MOC may take different forms, and traditional approaches should be challenged if they are proven to be sub-optimal to deliver the required balance between availability and cost. As a general principle, any chosen MOC shouldn't be reliant on specific venues for successful completion of maintenance periods, as this flexibility will provide increased confidence in the delivery of those maintenance periods.

In order to optimise the maintenance regime for a single platform such as that shown in Figure 4, a Class wide perspective needs to be considered as platforms will be drawing upon the same Support resources, i.e. people, facilities and infrastructure. Figure 5 depicts such a consideration for a fleet of 6 platforms, an indication of the Support resource demand at a given point in time can be obtained by taking a cut through the cycle for each platform as shown by the yellow dashed lines. Figure 5 illustrates the start of the maintenance cycle for each platform is driven by the build drumbeat, shown by 'x' in the chart. The value for 'x' may not always be consistent owing to build efficiency improvements, or delays in the build schedule for different platforms. If the maintenance cycle is not flexible, this could lead to stress on facilities and infrastructure and in turn a failure to meet the required platform availability. Furthermore, an enterprise view must also be factored in, as facilities may be shared with other defence assets including the succeeding class where any compatibilities exist.

Therefore, the focus in the design phase is to explore opportunities to de-conflict and de-risk maintenance periods. A number of options exist, including:

1. Improving the design of equipment and access arrangements optimise docking periods and ensure the dock related work can be delivered with confidence.
2. Optimise the docking duration so that they are less likely to overlap, through applying maintainability requirements to reduce work-in-way, and enables maintenance personnel to be more effective by, for example, facilitating concurrent working.
3. Enabling more in-water engineering through both Platform and Support Solution design to allow more work to be carried out afloat, primarily defect repairs and pre-emptive surveys prior to docking.
4. Providing assured confidence in the delivery of the maintenance periods by introducing a robust condition-based maintenance strategy, enabled by a Condition Monitoring capability as discussed later in this paper. Further to this point, development of a digital model to represent the physical condition of an asset enabling a robust prognostic capability that enables maintenance periods to be better planned

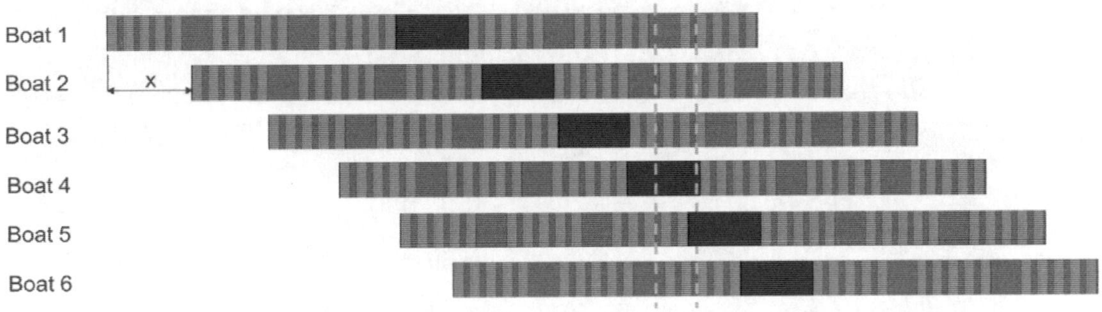

Figure 5: Maintenance and Operating Cycle – Class size and build schedule a consideration

4.3 INTEGRATED TEAMS

As the design progresses each of the decision-making elements needs to be considered as part of normal day-to-day design maturation. Making a balanced, high level singular decision can be unbalanced through normal design evolution if the key criteria used in the making of that balanced decision are lost. The most effective way of ensuring this sight is not lost is again to make each element owner a constituent part of the design team, with the appropriate stakeholder involved in the Programme from the very beginning [8]. With this in mind, the integrated teams should include designers, engineers, cost modellers, operators, builders, maintainers and equipment buyers. This would also need to include supporting services such as scaffolders, slingers and commissioning teams. Only with this breadth of input can each integrated team mature the design in a balanced way.

Care should be taken however to ensure that teams are empowered to make decisions in order to avoid a 'design by committee' which will likely introduce additional costs and could even inhibit design maturation as design decisions may not be made in a timely manner. Where requirements are at risk of non-compliance there needs to be a mechanism to revisit decisions or design progression on a case-by-case basis to ensure technical coherency and that the design is appropriately balanced and offers the optimal trade. A simple approach for managing this is through allocation of a weighting or 'exchange rate' between each of the highest level requirements to allow an objective trade-off to take place.

4.4 SUPPORTABLE DESIGN PRINCIPLES

Through stakeholder engagement with experts in the field of providing waterfront support to submarines, along with documented experience of past programmes, the Supportable Design Principles in Figure 6 have been identified that, if applied during design, will inherently lead to a more supportable platform. These need to be consistent with the Chief Defence Logistic Support (CDLS) 'Support Advantage Design Principles' which set guiding expectations for all projects across the life cycle [18]. This will ensure consistency with broader defence strategy and as such application of standards around Integrated Product Support [9] and Asset Management [19] is a compatible approach.

The advantage of these Supportable Design Principles is that, with coaching, they can be applied early on in concept before detailed requirements are established to build in desired Support characteristics. At early stages of the lifecycle the Supportable Design Principles are used as one of the main criteria sets for assessing the 'goodness' of the design from a supportability perspective. In developing these, it is recognised that some of the Supportable Design Principles may actually conflict with others. Like all complex designs it comes down to a trade-off with what is more important to the end user for a given decision. As the design evolves and requirements mature, the Supportable Design Principles are translated into useable and actionable design instructions that can be used at system and equipment level. These are equivalent to the Build Rules that are imposed on system designers and used to ensure that no anti-supportability design decisions are taken without valid justification and a clear understanding of the resulting risks.

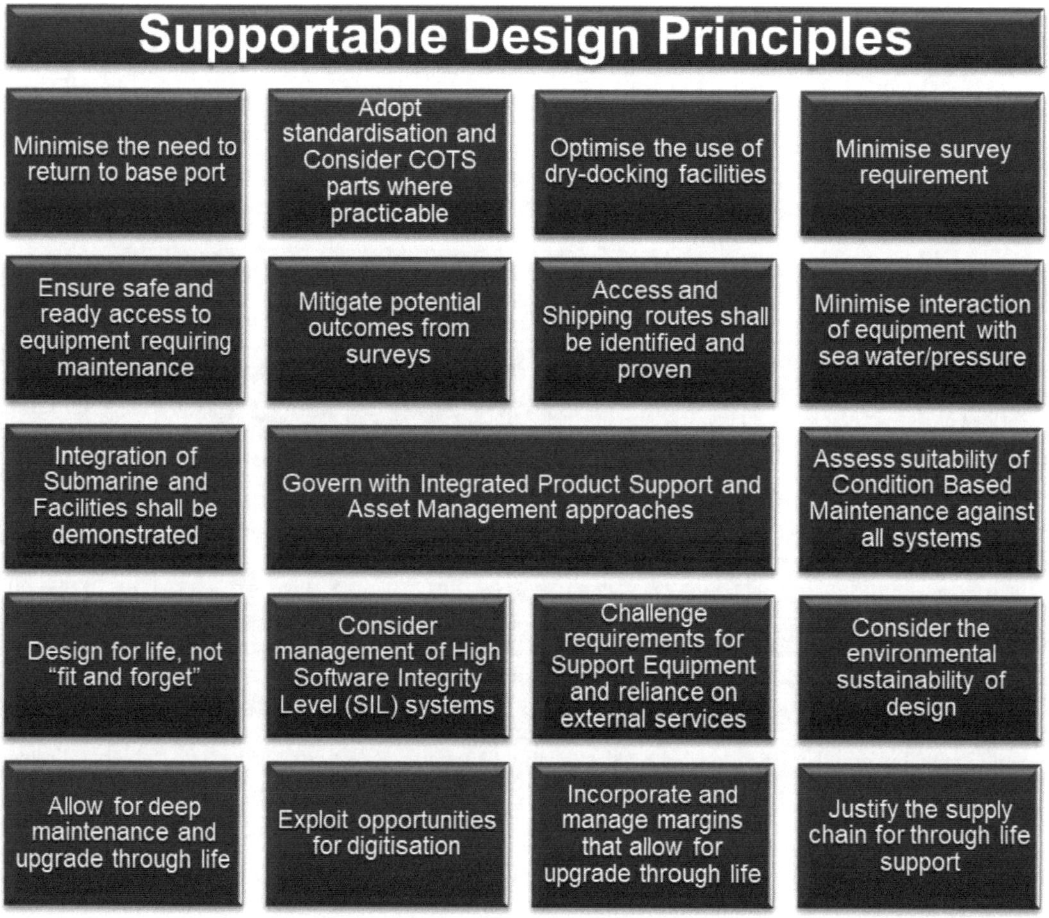

Figure 6: Supportable Design Principles

4.5 REQUIREMENTS AND RULES

There will be a suite of tangible availability requirements that have a defined measure of performance, design life and reliability for example. There will also be a series of rules that are more subjective in nature without a specific measure of performance. Without this measure they do not strictly qualify as requirements however they are just as important. For example, the inclusion of adequate maintenance envelopes and removal routes, appropriate placement of equipment within a compartment, Support needs within Technical Specifications for procured equipment and, spatial envelopes for support and test equipment amongst other things, all combine to derive a balanced and effective total system design [8].

Depending on the contracting framework between respective organisations and their customers, there needs to be a clear and configured link between the various requirements suites controlled by each Industrial Participants. Dependencies between requirement suites need to be established such that there can be no divergence between them. Incompatible and incoherent requirements will almost certainly result in a degradation to submarine availability. Using the build to in-service transition as an example, the Support Solution requirements need to be linked to the platform design requirements where appropriate and vice versa.

Where the design has been optimised for Support and availability has been safeguarded by design, then a simple data handshake can take place for the development of the related aspect of the Support Solution. Where the design is sub-optimal for Support and the risk to availability needs to be mitigated by the Support Solution then Support Solution requirements need to be derived to ensure this mitigation is coherent with the platform design and is effective. This is not a linear relationship however, top level requirements of the Support Solution that impact the platform design need to be linked to derived requirements placed on the platform design.

4.6 INTEGRATED PRODUCT SUPPORT

The complexity of 'Support' is internationally recognised through the development and introduction of various standards and specifications. Traditionally the UK has used Defence Standards to flow down requirements onto its contracted industrial base. Defence Standard 00-600 Integrated Logistic Support Requirements for MoD Projects is one such example of this. The Aerospace, Security and Defence Industries Association of Europe and the Aerospace Industries Association have published ASD SX000i [9]. This is a joint transatlantic specification where both American and European aerospace manufactures can participate to establish a complete and Integrated Product Support specification. The elements of Integrated Product Support can be seen in Figure 7.

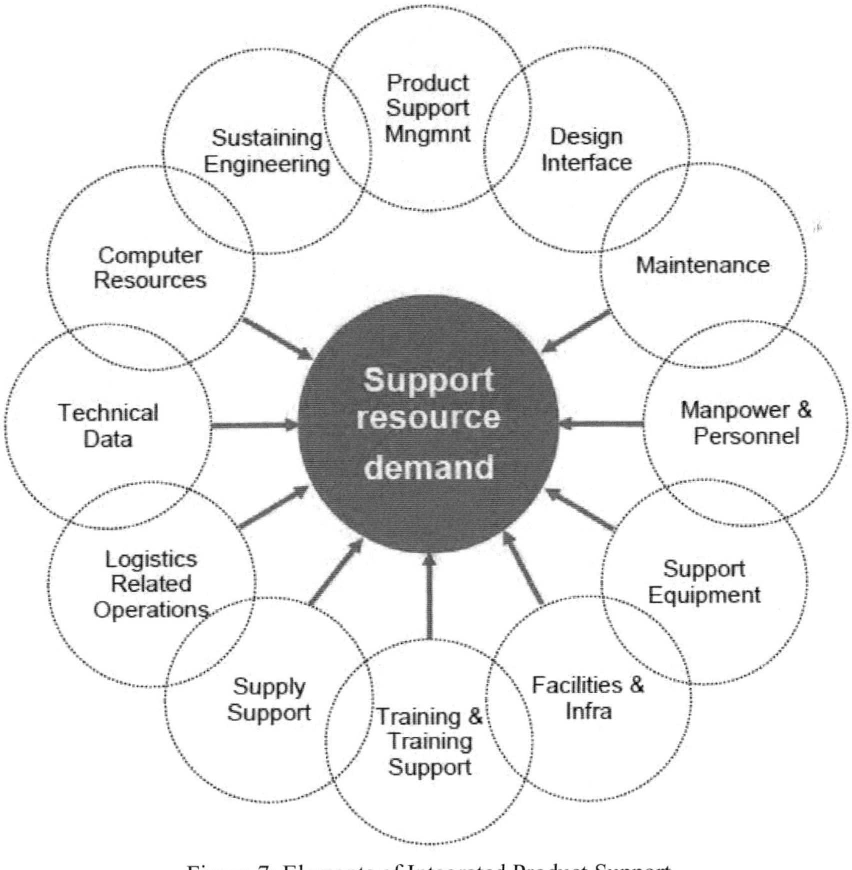

Figure 7: Elements of Integrated Product Support

Within the submarines Enterprise all Industrial Participants should sign up to and use the same standards, including the client. This will ensure that all elements of product support are properly captured, configured and contracted. This is essential to ensure there are no gaps in the development of the Support Solution and that the data required for this development is known early enough in the platform design phase that it can be requested from equipment suppliers with sufficient lead time to influence the design should any predicted Support shortfalls be identified.

In an environment where there is a continual pressure to achieve best value for money, a pragmatic and tailored approach to the application of the standards described is often required to strike the balance between Availability and Cost. This means that full adherence to the standards required may not always be appropriate but requires the programme to be 'risk aware' as removing or deferring aspects may result in gaps emerging in either design for Support optimisation or within the Support Solution itself.

4.7 ASSURANCE

Assurance to the end user that a product will achieve its specified function is critical to all businesses and their functions. The Assurance Case is a reasoned, auditable artefact that satisfies the claim that the product is good and technically coherent with the requirement set. The Case includes all supporting evidence, systemic arguments and explicit assumptions that support this claim. To that end a significant amount of time and effort (and subsequent cost) is sunk into this claim and the generation of the underpinning evidence, arguments and assumptions. Given the number of Industrial Participants involved across the lifecycle of submarines these costs can increase rapidly. A joint authority offers the best opportunity to minimise these costs by aligning the demands of the Assurance Case for each Industrial Participant. This alignment may be difficult to achieve initially and take time and effort to accomplish but the long-term benefit would be aligned industrial base and MOD governance processes such that Phase Reviews could be held in tandem with all Industrial Participants in attendance. This would have the added benefit of reducing the time taken to step through the assurance process of each Industrial Participant in turn. The Industrial Participants could also strive to achieve this alignment at the working level with a joint design authority to assess the design trade space between capability, cost and availability and direct the design according to the optimal benefit to the Programme.

4.8 LIFECYCLE TRANSITIONS

The management of lifecycle transitions within a single Industrial Participant can be easily managed as part of that Industrial Participant's own internal processes governing phase transitions. Phase transitions that cut across multiple Industrial Participants are more difficult to manage effectively. These transitions should therefore be undertaken over a long a period as possible to mitigate the risk of issues and shocks emerging after the transition has taken place. One way to do this would be for the Industrial Participants to share resource such that training opportunities can be established and explored. For example, in-service maintainers could be embedded into the build test and commissioning teams and with Ship's Staff maintainers to get first-hand experience of the planned maintenance activities. In-service maintainers could also witness design validation activities relating to the Support requirements and rules.

4.9 THE BALANCE

Only through careful and thoughtful integration of all Industrial Participants, stakeholders and interested parties into an endeavour with a single-minded focus on what is best for Enterprise can this balance be achieved. Additional time and effort in design could reduce the build drumbeat and improve the supportability of the submarine. Similarly build drumbeat optimisations may introduce significant supportability constraints. It is the striking of this balance that is essential in delivering a future submarine with the required capability and availability.

This balance extends into parity between Build and Support infrastructure and facilities. Accelerating the build drumbeat of submarines without increasing the capacity of the Support facilities will result in submarines starting to queue, waiting for facilities to become available to conduct planned maintenance. This planned maintenance could be deferred to keep the queue length down. However this carries an increased risk to the safety of the submarine and the submariners and to planned availability and capability from emerging defects, some of which may be more onerous to rectify than conducting the preventative maintenance as planned.
This Enterprise 'heartbeat' needs to be synchronised to ensure the existing facilities are prioritised and targeted for investment as required. The need for additional facilities needs to be identified as early as possible in the Programme to ensure that the 'required by' date is met bearing in mind these facilities need to be, in reverse chronological order; commissioned, tested, fitted out, built, designed, planned, budgeted, and have the necessarily political will to make happen.

Through the implementation of an asset management approach this balance can be realised as submarine availability. Using Integrated Product Support as the framework for requirements, rules and the collaborative venture from the outset, the design of the submarine and the Support solution can be optimised in harmony to deliver submarine capability and availability with agility and pace. The balance will provide 'Support Advantage' [10] to complement 'operational advantage' in enabling the Royal Navy to achieve the readiness levels required by Future Force 2020 shown in Figure 8.

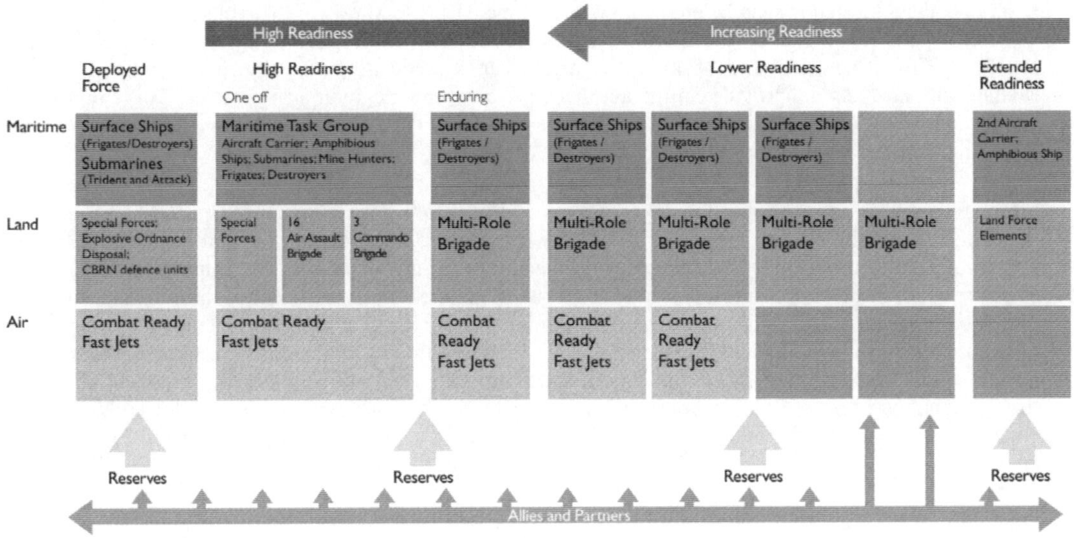

Figure 8: Future Force 2020 [11] © Crown Copyright 2010

4.10 KEY PERFORMANCE INDICATORS

Both Maintainers and Operators want, arguably above all else, certainty with regard to the requirement for maintenance, and the availability of the platform to deliver operational requirements. Even extending the periods within which maintenance can be delivered may be more preferable if it means that forecasting is more accurate, as this would allow deployed operations to be better planned and delivered [14].

Therefore, understanding the operational requirement so that the optimum maintenance cycle can be derived is critical. Success, from a Support perspective, is then measured on the confidence of delivering this maintenance and operating cycle once in the in-service phase of the lifecycle. As the programme progresses through the lifecycle, available data relating to the ability to forecast performance becomes more mature, and with that confidence in predicted performance increases. The level of confidence in delivering the Maintenance and Operating cycle at a given point in the lifecycle may drive action to change either the design or Support Solution, which includes the Maintenance and Operating Cycle, or take on more risk.

5. FUTURE CONSIDERATIONS

Technology is advancing at a considerable pace and with each step forward opportunities are emerging to improve the availability of submarines. This technological advancement also introduces supportability challenges as new and emerging capabilities are developed and inserted into the design of new submarines and retrofitted on current in-service Classes.

Not all emerging technologies are suitable for implementation onto submarines and any that are considered will need to be judged through the respective lenses of cost, capability and availability. With technological advancement also comes the increased risk of obsolescence, either through technological evolution or due to market forces; technological dead ends like VHS vs Betamax or Blu-ray vs HD DVD for example.

Each potential opportunity will need to be assessed in the same way that the submarine and Support Solution designs have been balanced to offer the cost-effective solution. It may be that some options are judged to be ineffective and therefore not developed for embodiment. Any change will also have to align with the Digital Strategy for Defence [12] to ensure the change can be exploited to the full. Technology and data incompatibilities within defence will lead to inefficiencies that result in a degradation of either capability, availability or both as well as escalating costs to force a compatibility fix.

One existing technology to focus on that has not yet been exploited to the full within submarines is Condition Monitoring (CM). CM will allow for the current Reliability Centred Maintenance strategy to be complemented by Condition Based Maintenance. CM could provide significant availability improvements provided the failure being monitored for can be predicted with sufficient time to allow for the replacement equipment to be supplied by the manufacturer and for the replacement activity to be planned into the maintenance period schedule. This prognostics and health management can also be used to validate the Reliability Centred Maintenance analysis and provide confidence that equipments are degrading as expected. This could allow for a reduced system survey requirement during maintenance periods as the monitoring system could provide the required confidence that the system is performing as expected.

Structure can also be monitored for degradation thereby reducing the demand for hull surveys. Submarine structures are usually coated with a protective barrier to reduce or halt corrosion and increase the life of the structure. These barriers need inspecting and where failures are identified, the impact on the underlying structure needs to be established and any remedial work planned and executed. Monitoring internal tanks for corrosion products is one such opportunity for minimising survey requirements in tanks for example. Only once corrosion products are identified in the tank will an inspection be planned. All forms of health monitoring need to be considered as part of the cost-benefit trade with availability as they may degrade capability through the consumption of network bandwidth margin or weight growth margin. Only through the collaboration and engagement of all Industrial Participants can this trade be fully comprehended and a truly balanced design result.

Digital models can also be created to provide additional evidence for the asset management decision making process when deploying a submarine and what the risk is to mission success. There are statistical limitations however with submarines as the limited number of assets feeding data into these models may not be adequate to draw statistically sound conclusions. Typically thousands of data entries for dozens of assets operating across various regions globally are required to give adequate confidence in the output of the digital model. This needs to be considered as part of the cost-benefit trade with availability before any effort is put into creating these models.

5.1 CONDITION MONITORING IMPLEMENTATION PRINCIPLES

From a supportability perspective, it is important to highlight that any CM solution should adhere to the following principles:

1. The CM solution should be sophisticated enough to facilitate planned maintenance and optimise operation in order to keep downtime to a minimum. It is critical to recognise the importance of building the 'digital twin' of equipment as early as possible, starting with factory testing data of equipment. This will set a baseline as to the sensing parameters equipment is expected to produce before it is integrated with the rest of the platform and operated in its intended context. The digital twin for equipment should mature with its design.
2. Caution does need to be exercised however when implementing CM in to the design. Notwithstanding the principle above, incorporation of simplification principles is key to achieving a platform that is capable, affordable and supportable. There may be a temptation to design an overly complex CM solution which inadvertently causes the user more problems than it solves. Revolutionary designs must offer a significant reduction in maintenance burden and/or cost, or address an issue with reliability of the previous method.
3. The strategy for managing data and corresponding condition based maintenance needs to be clearly thought through as part of any options analysis of a new CM solution.
   Condition Based Maintenance can provide great improvements in Availability, when deployed appropriately. Systems should be carefully assessed for potential Availability gains by application of Condition Based Maintenance. The assessment should consider;
   - Criticality of System,
   - indicators of functional failure of the items to be monitored,
   - Deterioration characteristics of the items,

- Ease and cost of other inspection methods.

4. Access to CM data is crucial. As alluded to in point 3 above, data relating to the operation of submarines attracts a security classification which becomes more restricted once aggregated and consequentially inhibits the ability of those responsible for maintaining the platform to access. Any design needs to provide the ability to harness available on-board data to inform decisions around Support and maintenance.

5.2 ADDITIVE LAYER MANUFACTURING

Another developing technology that has the potential to improve Support of Submarines is Additive Layer Manufacturing (ALM). ALM enables more efficient parts procurement, removing non-value adding links in the supply chain and reducing the likelihood of supply chain lead times impacting upon availability. This shortening of the supply chain, coupled with improved manufacturing efficiencies resulting in reduced wastage of material, leads to significant cost savings [17]. Furthermore, obsolescence risks can be significantly reduced by retaining appropriate design information such that required parts can be produced at the push of a button.

The UK governments NetZero commitment has led to a shift of the tide towards designing for sustainability, this is echoed in the CDLS Support Advantage Charter [10]. ALM offers a significant contribution in this regard through the recycling of materials, reduction in raw material wastage and less scrappage [17]. Before ALM can be introduced as a business norm for the Submarine supply chain, there are challenges that need to be overcome. These centre around repeatability of production, associated industrial standards and the capital investment required to set up the associated infrastructure. However, the potential benefits suggest that ALM is an opportunity well worth exploring for Submarines of the future that are being designed today.

In anticipation of this, there are considerations that can be made in design:

1. Space allocation for any ALM equipment and raw materials. While it is acknowledged that the conditions aboard a submarine don't lend themselves to the precise atmosphere required for ALM, carrying the raw materials and ALM machinery could prove useful if parts are required on deployment.
2. Component design such that ALM is a viable option for manufacture. This means both in the structures from which the component is formed, along with standardising material used across the design such that the same raw material can be used to produce a variety of parts.

6. CONCLUSION

In conclusion this paper has looked at some of the major considerations in the submarine availability and capability trade space. It has outlined how 'optimum asset value' can be achieved through balanced decision making for submarine availability, capability, together with build agility and pace through the close collaboration of the key Industrial Participants, acting as 'teammates' through a 'best for Enterprise' approach. It outlines how whole life considerations can be made through the introduction of design principles, framed through the adoption of an Integrated Product Support standard.

There is undoubtedly more that can be done to strike the best balance between capability and availability however this paper reflects the views of the authors and their intent with regards to Support integration on the Programme they are employed on.

7. REFERENCES

1. Government UK, 'Successor Submarine Recruitment', Website, 10 Jun 2013

2. Parliament.UK, 'Memorandum by the Ministry of Defence Readiness Assumptions', Website, Jun 2004

3. Parliament UK, 'Defence Sub-Committee, Oral evidence: Defence in the Arctic HC 388', Jan 2018

4. MoD UK, 'Joint Doctrine Publication 02', *UK Operations: the Defence Contribution to Resilience*, Nov 2021

5. BAE Systems, 'Astute Class Submarines', Website, May 2023

6. MoD UK, 'Defence in a Competitive Age', *Secretary of State for Defence*, Mar 2021

7. MILLS, C., & KIRK-WADE, E., 'The Cost of the UK's Nuclear Deterrent', *Research Brief*, May 2023

8. RAND, 'Lessons from the U.S. Navy's Ohio, Seawolf, and Virginia Submarine Programs', *Research Brief*, 2011

9. ASD-STAN, 'Setting the Standards for the Aerospace, Aviation & Defence Industry', Website, May 2023

10. StratCom UK, 'Defence Support Strategy Overview', *Defence Support*, April 2022

11. Government UK, 'Securing Britain in an Age of Uncertainty, Strategic Defence and Security Review', Oct 2010

12. MoD UK, 'Digital Strategy for Defence, Delivering the Digital Backbone', Apr 2021

13. CULLEN, D., 'Trouble Ahead, Nuclear Information Service', Apr 2019

14. CONGRESSIONAL BUDGET OFFICE, 'The Capacity of the Navy's Shipyards to Maintain Its Submarines', Mar 2021

15. 'Navy Lookout, Critical Royal Navy Submarine Refit Running Late', Website, August 2019

16. MUSSELWHITE, S., 'Methods to Reduce Backlogged Maintenance of Los Angeles Class Submarines', Apr 2021

17. JAVAID, M., 'Role of Additive Manufacturing Applications towards Environmental Sustainability', Oct 2021

18. MoD UK, 'Support Advantage Charter', Website, May 2022

19. ISO 55000:2014, 'Asset Management; Overview, Principles and Terminology', Website, May 2023

20. Royal Navy, 'AUKUS Deal Delivers New Class of Submarines for UK and Australia', Website, Mar 2023

MBSE DRIVING DOWN THROUGH LIFE COST IN NAVAL APPLICATIONS

I Tolchinsky and **A Davenport**, Ansys, USA

SUMMARY

Model Based Systems Engineering (MBSE) has become a well-established practice in the defense industrial base, especially for aerospace platforms, and is proven to lower life cycle costs. In this paper we explore the adoption and return on investment of MBSE practices in naval programs, specifically the US NAVY. MBSE methods applied during the acquisition, and throughout the life cycle, have made measurable reductions in cost for engineering, software development, and maintenance as well as providing a framework for implementing new technologies on existing platforms. The MBSE methodology enables the reduction of life costs as it provides a way to better manage the increasing complexity of naval platforms and weapons systems integration. Systems complexity and software integration are key issues in driving higher development and program support costs and leading to substantially longer development cycles. MBSE methods provide an authoritative source of truth (ASoT), accurately describing the system, the interactions within the system, and providing a mechanism to verify the behavior of the system and ensure the requirements are met. Specific examples from open-source material will be presented and analyzed to demonstrate how MBSE methods are delivering on the promise.

NOMENCLATURE

DCSE	Document-centric systems engineering
MBSE	Model-based systems engineering
pareto front	Set of optimal solutions in multi-objective optimization
PEO SUB	Program Executive Officer, Submarines
ROI	Return on investment
SE	Systems Engineering
SE&I	Systems Engineering and Integration
SEBoK	Guide to the Systems Engineering Body of Knowledge
SoS	System of Systems
SWFTS	Submarine Warfare Federated Tactical Systems

1. INTRODUCTION

Model Based Systems Engineering (MBSE) promises to improve the product development process by delivering better products faster and add at cheaper overall life cycle cost. This reputation caught the attention of the Submarine Warfare Federated Tactical Systems (SWFTS) program. In 2010 they put together a feasibility study[1] and initiated a transition from a document-centric systems engineering (DCSE) to an MBSE process.

2. APPLICATION OF MBSE TO THE SWFTS PROGRAM

SWFTS is a rapidly evolving combat system of systems (SoS) product family. The SWFTS architecture supports a common product line across seven submarine classes with a total of ten platform variants, where each platform variant contains a similar set of component systems and information architecture with variations due to platform class, age, the specific sensors and weapons launchers employed, and the mission capabilities delivered. A diagram of the architecture is shown in Figure 1.

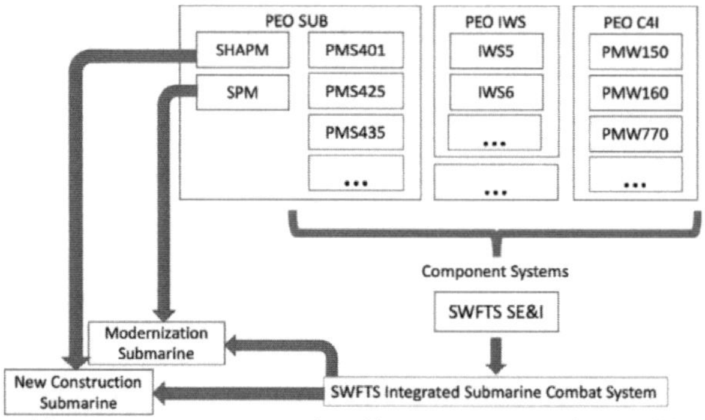

Figure 1: SWFTS architecture supports a product line across 10 variants with shared components.

To test out the application of MBSE, the SWIFTS Systems Engineering and Integration (SE&I) program put together a case study. The objective of this case study was to demonstrate the effectiveness of and document the Return on Investment (ROI) achieved by applying an MBSE methodology to the SWFTS SE&I program. This was at a time when there was little evidence in the literature for a positive ROI for moving from DCSE to MBSE.

In the context of the SWFTS SE&I program, MBSE is the application of an integrated model-based systems engineering process using SysML and UML23 for the management of requirements, architecture, design, analysis, verification and validation activities in support of the on-going evolution of a submarine combat system product family. The case study focused on managing the requirements on the interfaces between component systems.

The results of the study exceeded all expectations.

The MBSE approach was more efficient. The team found that the SE&I systems engineering effort to modify and configuration-manage interface requirements in a large system of systems is 18% more efficient using an MBSE approach, as compared to a requirements database approach.

Interface defects are discovered earlier using MBSE. 18% of interface defects were discovered earlier in the life cycle, where they are much cheaper to remove. Defects discovered during platform and acceptance test cost four to fifty times as much to correct as defects discovered during integration.

The conclusions are summarised in Figure 2.

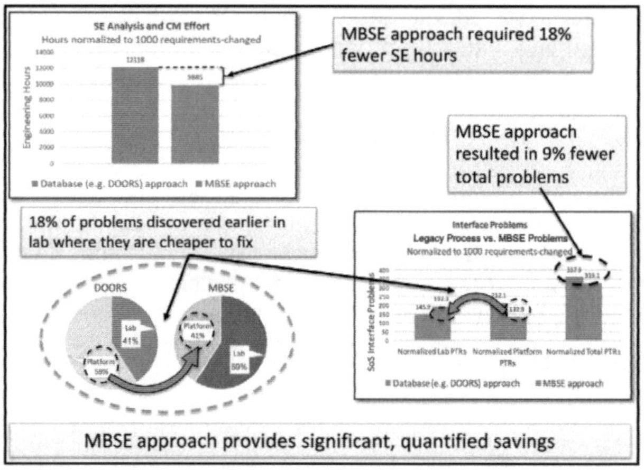

Figure 2: Summary of the benefits of the MBSE approach.

MBSE produces a higher quality product. There was a 9% reduction in total defects. This has a great cost implication as defects that escape into the Fleet cost up to an order of magnitude more to correct when they are discovered after deployment.

The MBSE implementation provided many other benefits that are more difficult to quantify.

The model provides a centrally accessible source for SoS architecture and design that improves the capability of the SE&I team to manage SoS complexity. The model also facilitates a better shared technical understanding among the collaborators. It provides the necessary traceability between capability, requirement, function and deployment – capturing details often lost to tribal knowledge when answering "why/what/when/who/how?" regarding implementations of complex SoS designs.

The case study was able to provide a rare precise ROI. Figure 3 shows a summary of the ROI for the SWFTS MBSE transition.

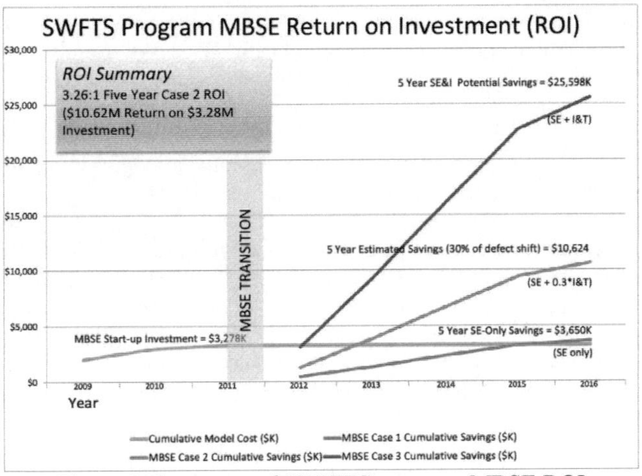

Figure 3: Summary of SWIFTS program MBSE ROI.

When normalized to a notional 1000-requirements SWFTS update, these MBSE-driven savings are between $334K and $2,342K, excluding the initial investment in the MBSE transition. Based on nominal development metrics, the customer organization, the U.S. Navy Program Executive Officer, Submarines (PEO SUB), saved between $2.56 M and $6.19 M on the MBSE Baseline update using the MBSE method, which is comparable with the projected savings of $900K, from an earlier pilot study.

The transition from DCSE to MBSE has been warmly received by stakeholders across the submarine combat system enterprise. Since the traditional SWFTS SE&I work scope is being accomplished for less cost, the customer has expanded the scope of Systems Engineering (SE) work being accomplished within their budget. Suppliers are getting better quality SE products as inputs to their work, which has led them to adopt MBSE for their own system maintenance and evolution programs. The positive ROI achieved by the SWFTS MBSE transition has encouraged both legacy and new-start programs across Lockheed Martin to adopt MBSE. Other parts of the U.S. DoD acquisition community have taken note and are using the successful SWFTS MBSE transition as evidence to support their own plans for digital engineering transformation.

3. SEVEN SE PRINCIPLES RELEVANT TO THIS APPLICATION

The Guide to the Systems Engineering Body of Knowledge (SEBoK)[2] identifies seven SE principles that provide insight on the benefits realized in the case study. These are:

Requirements traceability LP1: MBSE improves traceability in multiple dimensions but maintaining requirements traceability between a traditional database and the MBSE system model can be challenging. While DOORS, MagicDraw and Teamwork can interoperate to provide requirements management and traceability, the combination is fragile. Without careful configuration management, synchronization can lead to database corruption. If DOORS and Teamwork are on separate servers, maintaining the connection can run afoul of ever-evolving corporate information assurance (IA) policies.

Requirements can be managed using the SysML language inside the system model quite effectively. This approach can reduce the resources needed to keep the system model in sync with a traditional requirements database system and increase overall SE productivity.

Communications LP2: Tailored SE products generated from the system model can substantially enhance communications both within the technical team and between customer stakeholders. Graphical depictions of the system model often communicate better to human stakeholders than massive spreadsheets and textual documents. Further, the enhanced precision driven by modeling can reduce miscommunications between both technical and programmatic stakeholders.

Having the architecture and design in a system model makes it affordable to generate specialized SE products on demand for particular communications needs while keeping all SE products in concordance. Even technical stakeholders who thought they understood the design can find new insights by looking at it in different representations.

Productivity LP3: MBSE increases productivity by enhancing communications within the team, automation of routine tasks, and cost avoidance. Processes used in DBSE often require substantial revision to achieve the potential productivity gains of MBSE. In particular, review processes must be modified to take advantage of the tooling.

The selected modeling tools constrain how you can practically re-engineer SE processes. Automation can replace a great deal of routine SE work (document generation, identifying potential impacts of changes, etc.).

Developing strong modeling style guidelines and specialized representations, along with training materials to indoctrinate new team members as they join the program, is worth the investment. MBSE does require a trained cadre of modelers, but not all systems engineers have to become skilled modelers.

To effectively quantify the benefits of MBSE, a program needs to plan metrics collection carefully, and then stick to the plan long enough to collect meaningful data.

Quality LP4: Much of the ROI from the MBSE transition can be in improved quality. Improving the quality of SE products reduces latent defects in systems delivered to the customer, reducing maintenance costs and increasing customer satisfaction.

Models are less tolerant of imprecision than documents. The increased precision improves SE product quality, both in reduced defect generation and in reduced defect escape. The automation of product generation can make specialized SE products affordable, further enhancing system quality.

Managing change LP5: How a proposed system change is understood and executed is fundamentally different between a model- and a document-centric approach. In the document-centric approach, the focus is on "What should my final output look like?" In the model-based approach, the focus is on "What does this change mean to the system? Which other parts of the system are impacted by this change?"

Change management is hard. When moving from DBSE to MBSE the approach must be carefully considered up front, and the system model must be designed to facilitate that approach. Change management also impacts tool selection, since different tools align better with different approaches.

Managing variants LP6: The most common process for managing variants in DBSE is 'clone and own', where each new product family member takes the then-current baseline and 'forks' the baseline for evolution of the variant. This makes synchronizing changes to the core baseline across the product family a very labor-intensive process. Treating variants as deviations from a core baseline in a model greatly reduces the cost of managing variation in a product family.

Variant management is difficult. The planned approach must be considered up front, and the system model must be designed to accommodate it. The selected approach impacts tool selection and tailoring.

Design the system model to treat variants as deviations from the core baseline. Then changes to the core baseline are automatically shared among all variants, and impact to product family members is limited to any impact of core baseline change on specific variant deviations. This also facilitates commonality between variants, a key customer goal as commonality reduces logistics and training costs.

Life cycle LP7: MBSE can be applied early or late in the product family life cycle. While most projects using MBSE start off model-based, a program can transition to MBSE late in the life cycle.

4. FURTHER GAINS

The positive ROI achieved by the SWFTS MBSE transition demonstrates that MBSE is not just for new-start programs and should encourage other programs and organizations in their consideration of moving from DCSE to MBSE. The study also notes that the observed ROI should be considered a lower bound, that there is more to gain from MBSE.

Further benefits from adopting the MBSE approach can be obtained by connecting the system architecture models to simulation. The descriptive system models become executable. ModelCenter from Ansys provides a unique capability in this domain. It can take the engineering tools already used by an organization to evaluate the performance of designs, automate their execution and provide a bi-directional link with the system model. This is visualized in Figure 4.

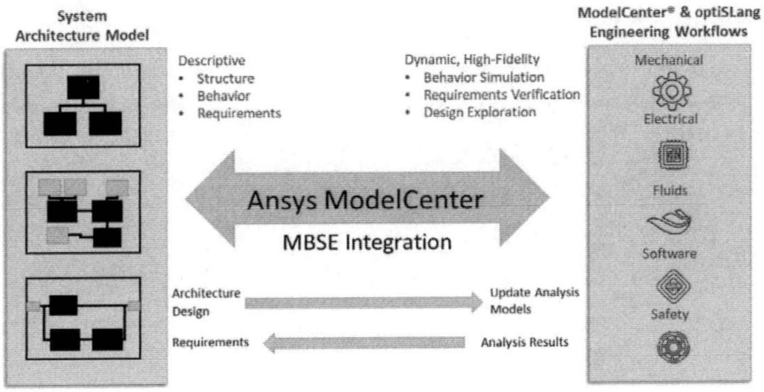

Figure 4: Bi-directional link between SE and Design.

Systems engineers can evaluate the performance of a particular configuration and see if system meets its target requirements. In Figure 5, value properties in the system model are mapped to corresponding requirements.

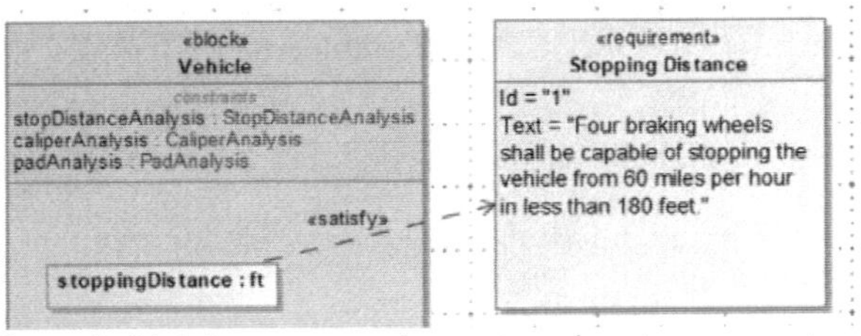

Figure 5: Link between requirements and system parameters.

This information is processed by ModelCenter to provide feedback on the state of the requirements as shown in Figure 6.

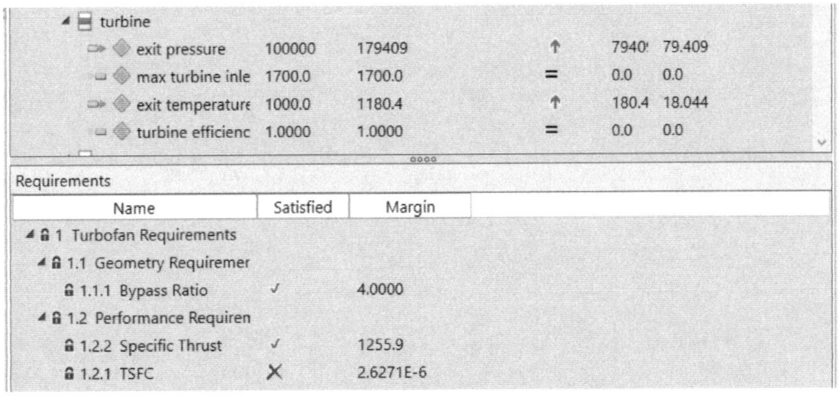

Figure 6: Requirements status summary.

The real power of ModelCenter comes into play when applying its Design Exploration capabilities. The systems engineer can now run trade studies to gain insight into the behaviour of the system. It is possible to evaluate the performance of all possible alternatives and choose the best one.

Figure 7 shows how Airbus Defence & Space has used ModelCenter.

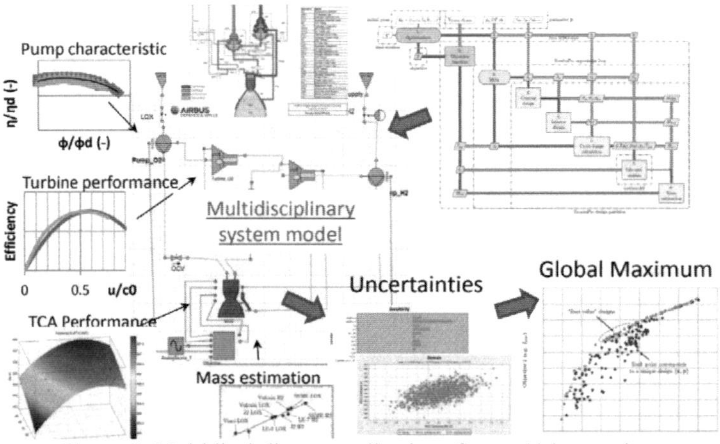

Figure 7: Multidisciplinary applications for an Airbus project.

When considering the physics of the individual components at the system level it is possible to understand early on which design disciplines will have a strong interaction that will impact the overall system design.

By understanding the relationships between competing performance objectives, it is possible to identify when the best trade-offs occur and place the product into this sweet spot.

The expected impact of manufacturing tolerance uncertainty can be calculated and mitigated.

Being able to do this at the system engineering level enables the enterprise to be on the right path from the start and avoid costly late course adjustments.

5. ENGINEERING AUTOMATION IN COMBAT SHIP DESIGN

A Model Based Environment is today commonplace in all engineering applications including the combat ship arena. CSRA Inc. uses ModelCenter to perform hull form design studies[3]. They are able to automate the execution of a number of design tools including Rhino 5 with Orca3D for parametric hull form generation, Simerics for high fidelity CFD, and SMP, a seakeeping analysis tool. The ModelCenter workflow that democratizes the execution of these tools is shown in Figure 7.

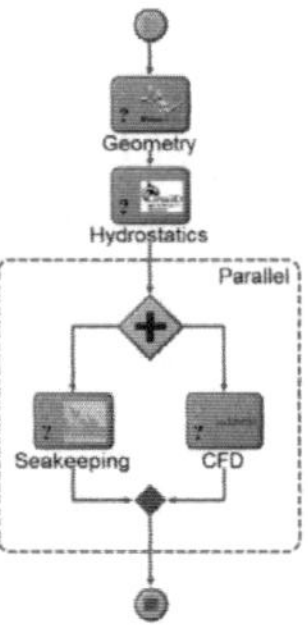

Figure 8: ModelCenter workflow for the hull form analysis.

Figure 9 shows the hull baseline and the optimized result. Baseline is black and the optimized hull form is orange.

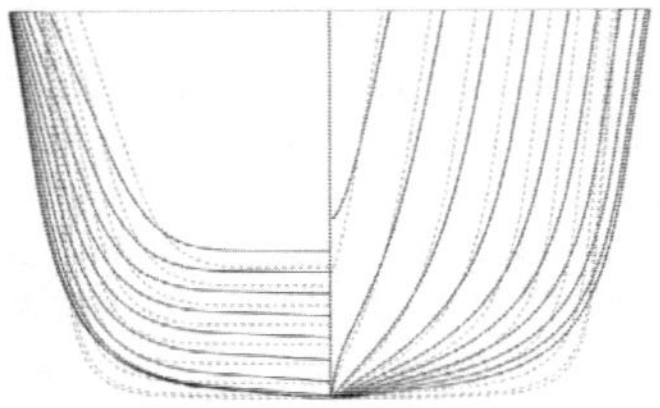

Figure 9: Baseline and optimized hull form.

The team was able to run thousands of design points and gain valuable insight into what was possible in terms of trade-offs in performance. The sampled data along with a 3D pareto front are visualized in Figure 10.

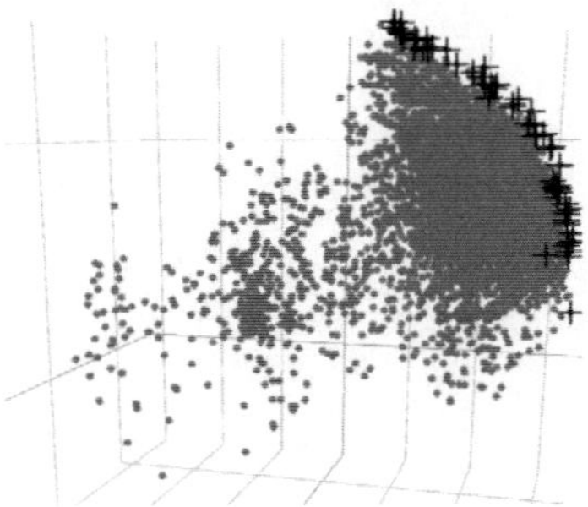

Figure 10: Design points evaluated during an optimization run.

6. CONCLUSIONS

The SWFTS SE&I program was able to clearly demonstrate quantitative and qualitative improvements in their SE process through the adoption of the MBSE approach. The greatest impact to life cycle cost was the reduction in defects that would be very expensive to address after deployment.

Further benefits can be gained by connecting the system architecture models to simulation. This provides the capability to verify that the system will meet its target objectives. The integrated workflow can provide insight into what is possible for a candidate architecture by performing system-level performance trade studies.

7. ACKNOWLEDGEMENTS

The authors would like to express their gratitude for the work of the authors on whose work this paper builds. They would also like to thank all the engineers with whom they have collaborated to move this technology forward.

8. REFERENCES

1. ROGERS III, E.B., MITCHELL S.W., 'MBSE delivers significant return on investment in evolutionary development of complex SoS', *Syst Eng.*, 2021;24:385-408.

2. 'Submarine Warfare Federated Tactical Systems', The Guide to the Systems Engineering Body of Knowledge (SEBoK), v. 2.7, 2022

3. ADAMS, J., Mizine, I., Rogers, C., 'Phx2018UC_MDAO_CSRA_Adams-Mizine-Rogers.pdf', Available at: https://www.phoenix-int.com/wp-content/uploads/2018/05/Phx2018UC_MDAO_CSRA_Adams-Mizine-Rogers.pdf, 2018

9. AUTHORS BIOGRAPHY

Ilya Tolchinsky holds the position of Lead Application Engineer at Ansys. He has an aerospace background and has been involved with engineering design software for the past 25 years. Today he focuses on supporting Ansys activities in the Model Based Engineering environment.

Anthony Davenport is the director of the North American Systems team at Ansys. He has spent over 30 years focused on working with organizations to perform "smart" digital transformation. Most recently, the integration of system engineering with mission, physics and cost models has enable new technologies that allow organizations to connect with their suppliers and their customers for making complex product decisions and win more projects. Mr Davenport and his team of system engineering experts at Ansys are helping customer put open architectures in place that allow them to exceed their organizations goals and connect system of systems for all parts of the product life-cycle.

Warship Conference
31st May – 1st June 2023
Guildhall, Bath, United Kingdom

INCLUSIVE SUBMARINE DESIGN CONSIDERATIONS

J A Dutton, BMT Canada, Canada

SUMMARY

Today's submarines need to accommodate for a variety of sailors. Submarine designs need to account for differences across a wide range of sailor size, height, body type and gender all while supporting optimal operational performance. This demographic shift demands re-evaluating elements of traditional submarine design and operation to optimize crew performance and attract, accommodate, and retain the most qualified sailors regardless of physical or social characteristics. This paper will present inclusive submarine design considerations from both anthropometric and social perspectives, present some examples, and discuss some potential solutions for addressing these design challenges.

1. INTRODUCTION

Although submarines have been designed and crewed for over 250 years, it is only relatively recently that the presence of mixed-gender submarine crews have brought to light social and anthropometric deficiencies in traditional submarine design. Increasingly for today's submarine crews, both of those perspectives must be considered and integrated into submarine design if submarine operation, maintenance, and sustainability are to be optimized.

Since Norway began incorporating female submariners in 1985, an increasing number of countries are utilizing mixed-gender crews on their submarines as well. Today there are nearly 20 countries known to include women serving onboard submarines and include: Norway, Denmark, Sweden, Australia, Spain, Canada, Germany, Poland, South Africa, United States, Venezuela, Argentina, United Kingdom, Greece, Italy, France, Portugal, Netherlands, and Japan. It must be noted that in some cases (e.g., USA and UK), female submariners are still limited to their respective country's largest submarines (i.e., ballistic missile submarines) as increased volume permits greater flexibility with living arrangements.

While the increasing presence of mixed-gender crews has served as an impetus to revisit submarine design philosophy and practices globally, as a design community we should take the opportunity for the discussion to reach beyond gender considerations and include wider social and anthropometric aspects as well. While not an all-encompassing list, this paper considers some common examples and sample solutions from both social/psychological and anthropometric/physical perspectives for inclusive submarine design.

The opinions expressed in this paper are solely those of the author.

2. INCLUSIVE SUBMARINE SOCIAL/PSYHOLOGICAL DESIGN CONSIDERATIONS

2.1 WASHROOM FACILITIES (HEADS)

Submarine washroom facilities (commonly referred to as "heads" on naval vessels) are perhaps the most obvious area of interest when it comes to considering inclusive social design considerations. Heads typically include not only toilets and sinks, but (very cramped) showers as well (see Figure 1).

Figure 1: Typical Submarine Shower

Heads are often shared by large numbers of crew, who are usually on a tight timeline to prepare for their next event whether it be maintenance, assuming the watch, or simply to get out of the way to permit other sailors to utilize the facilities to meet their own strict schedules. The necessity of passing many personnel quickly through limited washroom facilities can complicate potential simple solutions for resolving privacy issues, such as taking turns individually or in groups.

It can be seen in Figure 1 that if a sailor brings a change of clothes into the tight confines of a traditional submarine shower to maintain privacy, the clothing will become damp at best but more than likely be thoroughly soaked by the end of the shower. Beyond time-sharing (setting aside separate dedicated time and space for men and women separately, often using signs to indicate current usage), some potential solutions include (but are not limited to):

- Changing clothes elsewhere (which would presumably also require privacy protection in that location).
- Modifying shower design to include a watertight compartment for clothing storage and increasing internal volume to permit room to dress and undress.
- Designing shower entry/egress as private spaces, instead of access directly from the general access area in the washroom.

2.2 BERTHING

On naval ships, the sleeping areas are typically referred to as "berthing" and beds are called "racks". Like most aspects of submarine living arrangements, berthing spaces are compressed and usually offer little in the form of privacy (see Figure 2).

Figure 2: Typical Submarine Berthing

Although berthing inclusivity has some of the same challenges as previously discussed for heads, there is often more flexibility to address those challenges with logistical arrangements (e.g., changing clothes in a dedicated private space, arranging for a single sex berthing room or section, etc.).as berthing is not usually the resource timesharing bottleneck that heads are.

Regardless of whom is using the racks, best practices are that blackout curtains be used and that they can be clasped and/or sealed in order to prevent curtains sliding open when the submarine inevitably takes angles or rolls. Additionally, it cannot be over-emphasized that submariners often view the limited space in their racks as the only slice of the boat that is truly "theirs" and the one place they can be alone, as submariners live at their work around the clock for weeks (or even several months) at a time. It is not unusual for sailors to decorate the inside of their racks with pictures and letters from home, for example. Therefore, in addition to the obvious function of keeping light out while sleeping having blackout curtains that can be secured often serve as an important psychological barrier between the only "me" space submarine crews are afforded for weeks or months on end.

2.3 OPERATOR WATCHSTATIONS

Operator watchstations serve as points where sailors can interface with various submarine systems to manipulate systems, monitor and analyse sensor data, or both. For the majority of submarine history, watchstations were fairly rigid and unable to be changed from operator to operator. This forced sailors to conform both their physical and mental natural processes/preferences to the watchstation, with corresponding learning curves and losses of operational efficiency. This in part was due to the inflexibility of the largely mechanical nature of many legacy watchstations, but early electronic watchstations had little or no reconfigurability as well.

With the rise of flexible software in the last 30-40 years, electronic watchstations are becoming increasingly configurable to suit operator preferences or optimize layout for specific operational scenarios. This trend shows every sign of continuing, supported by extensive research into Virtual Reality (VR) and Augmented Reality (AR) applications.

The ability of operators to configure their watchstation to suit their preferences can lessen learning curves and improve operator efficiency. However, individual operator preference must be balanced with having critical functions remain the same across operators to permit supervisors looking over their shoulders to process critical information quickly no matter who is on watch. Balancing reconfigurability for personal preference with the use of standardization for other stakeholders can be a tricky balance to achieve. VR and AR offer some potential solutions, e.g., watchstanders and supervisors could configure layouts and/or controls individually to their taste without impacting the way that others consume information. Minimally, best practices dictate that display and control layouts should be able to be saved on an individual basis and be able to be switched quickly between operators.

Although not covered here as a topic for inclusivity, it should be noted that watchstations that leverage the advantages and avoid the disadvantages of human multimodal processing will assist in optimizing watchstander performance.

2.4 DINING (MESS) FACILITIES

The limited living space in submarines often necessitates that spaces serve more than one purpose. For example, a countermeasure launching space can also serve as a nurse's office or a wardroom table mainly used for meals can double as an emergency operating table. This is perhaps nowhere more evident than in the crew's dining facilities (often referred to as "the mess") as the dining area is usually the largest gathering area onboard (see Figure 3).

Figure 3: Typical Crew's Mess

Other than the main function of providing a place for the crew to be served and have meals, some of the uses of the crew's mess include (but certainly are not limited to):

- Watching movies together;
- Playing games (e.g., cards, backgammon, etc.);
- Gathering (muster) point for events, such as Damage Control situations;
- Group training, briefs, meetings, religious services; and
- Individual use for studying or working.

Given that many of the uses for the mess are for crew relaxation, it should be furnished and decorated comfortably and in a manner that sets it apart from workspaces. As stated previously submarine crews live where they work around the clock, sometimes for several months. Given this, it is vitally important that the crew be provided with avenues to "get away" from work occasionally.

2.5 ENTERTAINMENT MEDIA

As previously discussed, providing for crew relaxation, and giving them a break from their work routines is important. Another example of a means to provide welcome distractions comes in the form of entertainment media.

40 years ago, entertainment media was nearly exclusively limited to a small "ship's library" (often a small locker containing publications donated by the crew) and perhaps reel-to-reel movies (provided by the Navy) shown in the crew's mess, necessitating very limited inventories due to space constraints. However, technology has enabled an exponential increase in the availability of entertainment in the form of digital media. While it is still common to show movies to a group (generally on the mess decks) for those desiring social events, it is becoming increasingly common to provide access to a ship's digital library from other locations such as directly from a sailor's rack in berthing for those who are seeking solitary activities or those that simply prefer another movie. To support the ever-increasing diversity of today's submarines, best practices suggest

that providing a wide selection of digital media across multiple genres to appeal across a wide range of audiences is a wise approach.

3. INCLUSIVE SUBMARINE ANTHROPOMETRIC/PHYSICAL DESIGN CONSIDERATIONS

3.1 GENERAL ANTHROPOMETRICS

Anthropometrics is the study and collection of data relating to measurements of the human body, and goes beyond simple height measurements. For example, width across shoulders or hips, or arm and leg length can vary widely even among members of the same gender much less across genders.

Over the last several decades, designs incorporating anthropometrics have often targeted their design range to the 5th percentile (5th%) female – 95th percentile (95th%) male in populations to include a majority of users. However, in the past few years the approach has shifted slightly to designing for "the central 90%". This shift in perspective from "designing for the extremes" is not only a more inclusive mind set of ensuring that 90% of the population is accounted for in the design, but also acknowledges that rather than a single set of anthropometric parameters (e.g., height) there are often several parameters that combine to form design requirements. In summary, the central 90% approach seeks to include 90% of the population after all anthropometric requirements have been considered (regardless of the statistical distribution) rather than considering a single set of anthropometric data and assuming a normal distribution.

Strictly from an anthropometric inclusivity standpoint, designing to accommodate more of the population over less is obviously preferred. For example, striving for the central 94% vice the central 90%. However, there may be other factors driving design that act as a constraint on the ability to account for more of the population as well as a point of diminishing returns. Therefore, in design it is appropriate to treat anthropometric fit as a risk: weigh the probability and consequence of any particular risk being realized and becoming an issue against the cost of mitigating that risk to an acceptable level.

Some specific examples of submarine anthropometric challenges follow in the remainder of this section.

3.2 WASHROOM FACILITIES (HEADS)

For improvements in inclusive design of shipboard heads, we can look to some of the improvements made for gender-neutral civilian bathrooms:

- Toilet height and size should be considered. Although toilets tend to be fairly standard internationally, their effective height can sometimes be altered when mounted in submarines (e.g., when placed upon system components such as macerators).
- All toilet stalls should include sanitary bins placed out of the way of the other components in the stall.
- Shower length, width, and height should be considered.

3.3 BERTHING

Racks are the central feature of a berthing space (see Figure 2), and obviously are critical not just psychologically as stated previously but physically as well. Some potential design considerations related to racks are:

- Ease of access for top racks. It is common for submarine racks to be stacked three-high, resulting in a potentially difficult (and perhaps dangerous) climb for smaller sailors. The probability of danger is further increased if the ship is experiencing rolls and/or pitch angles. Footholds and handholds should be provided, and tested over the target anthropometric design range.
- Adjustable ambient environment. Sailors should have the ability to adjust the ambient environment (e.g., lighting, temperature, ventilation) in the rack to personal preference. For lighting control mechanisms rotary switches are often a better solution than switches, since light intensity can be adjusted gradually and activating the wrong switch doesn't unexpectedly and instantaneously provide a step change in light level.
- Sufficient interior dimensions. Some (but not all) of the factors to consider:
    - Space is needed above one's head (with pillow space factored in).
    - Sufficient space around shoulders is a key constraint.
    - When bodies relax, feet will flop thereby increasing the volume needed for foot space.

- Sufficient body space needs to be provided to enable sailors to roll over comfortably.

3.4 OPERATOR WATCHSTATIONS

As in the case of heads, we can evaluate design considerations from civilian examples to suggest potential design improvements for operator watchstations to support watchstander inclusivity and enhance comfort. In general, good workstation ergonomics still apply. For example, providing adjustable chairs, monitors, keyboards, mouse/trackball heights, and possibly footrests allow for easy customization to support individual anthropometric differences. Consideration may also be given to enable watchstanders to stand when desired, instead of forced sitting for hours at a time. However, submarine motion or angles may necessitate providing additional support to watchstanders while standing to ensure operator stability and safety.

3.5 CRITICAL/KEY/EMERGENCY COMPONENTS

A submarine is replete with components that not only require the ability to be operated on a day-to-day basis, but also quickly in emergency situations to prevent personnel injury, system damage, or even loss of the ship. Some examples include (but are not limited to): emergency isolation valves, hatches, steering and diving controls, some damage control gear, etc. The various actuating mechanisms historically have been numerous and required various strength levels to operate (see Figure 4).

Figure 4: WW1 German U-boat

While many critical, time-sensitive components are operated with powered assistance (e.g., hydraulic, electrical, etc.), if the component is truly vital for safety of ship there will normally be a backup mode (or even several modes) of operation that require manual operation.

Given that manual (unpowered) operation is typically supported for all critical components, designers should consider both ease of personnel access and ease of operation of these items. Access alone isn't sufficient if the sailor doesn't have the strength to operate the mechanism once they reach it.

Although this section focused on critical components due to safety implications, ease of access and operation should be considered by designers for even those components that aren't safety critical to support efficient operation of the submarine across the wide range of human forms. For example, an oil sample tube that requires sailors to reach around gear and operate valves unseen is at minimum inefficient and potentially dangerous.

3.6 CLOTHING AND DAMAGE CONTROL GEAR

Certain articles of clothing and Damage Control gear are provided by the ship, and as such should fit the crew for personnel safety, efficiency, and comfort. Examples of ship-provided clothing include (but may not be limited to):

- Anti-Contamination ("Anti-C's") clothing for radiation workers (on nuclear vessels).
- Maintenance/cleaning coveralls.

Quick and clear access to emergency clothing and Damage Control gear should be provided across the range of body types with easily opened storage. Some examples of Damage Control gear that are ship-provided include (but may not be limited to):

- Flame retardant coveralls for firefighting.
- Portable air-breathing equipment, such as an Oxygen Breathing Apparatus (OBA) or Self-Contained Breathing Apparatus (SCBA).
- Emergency Air Breathers (EABs). These items are not self-contained, as they require plugging an air hose to air supply manifolds located throughout the ship. Therefore, not only is it important to consider quantity, size, and ease of access to the EABs themselves consideration must also be given to the air supply manifolds. Air supply manifolds should be:
    - Of sufficient quantity to support not only the initial Damage Control team combatting the casualty, but the needs of all the crew if air-breathing assistance is required long term.
    - Located wherever crew are expected to be to operate the ship (both in normal and emergency modes) and carry out their routines. For example, while it is necessary to locate manifolds on watchstations wherever the operator is expected to travel and operate equipment, it is also essential that there are manifolds in sufficient quantity in berthing, the mess decks, the heads, passageways, etc.

4. CONCLUSIONS

After greater than two centuries of submarine design and operation based on generic "standard" males, the last few decades of incorporating mixed-gender submarine crews have emphasized social and anthropometric deficiencies in traditional submarine design.

Although mixed-gender crews have recently highlighted challenges to traditional submarine design and operation philosophy globally, as designers we should take the opportunity to broaden our thinking even further and recognize we must consider wider social and anthropometric aspects to enable safe and optimal operation of the submarine and to better retain submariners.

This paper presented just a few examples and potential solutions towards improving inclusivity in submarine design. There are many more examples and challenges as the submarine force becomes more inclusive globally, but history has shown that submarine designers will meet any challenges brilliantly.

5. ACKNOWLEDGEMENTS

The author gratefully acknowledges the generous and expert assistance of Chris Greenbank, Principal Human Factors Consultant with BMT Australia, in the development and review of this paper.

6. AUTHORS BIOGRAPHY

Allen Dutton currently holds the position of Combat & Systems Engineering Manager at BMT Canada. He is responsible for managing a team of over 24 Systems Engineering, Cyber Assurance, and former military operator professionals. While at BMT Canada, he has contributed to Canadian naval training and both Canadian and international submarine procurement projects. His previous experience includes over 22 years in the US Navy as a submariner on five submarines and one aircraft carrier, as well as five shipyard tours including new construction and overhaul of both submarines and aircraft carriers. Since retiring from the US Navy in 2006, he has participated in submarine and surface ship procurement and design in the United States, Australia, and currently in Canada. He holds a Bachelor of Science in Nuclear Engineering from Oregon State University (USA), as well as a Master of Science in Modelling, Virtual Environments, and Simulation (with work in Human Factors) from Naval Postgraduate School (USA).

THE EFFECT OF ORTHOTROPIC STEEL ON PRESSURE HULL COLLAPSE

J J Reijmers and **A Zambon,** Nevesbu b.v., The Netherlands

SUMMARY

Modelling the material properties of pressure hulls represents a crucial part in the design of submarines, as the operational profile and the maximum diving depth of these vessels tightly depend on the structural resistance of the hull to the external pressure. This paper investigates the application of orthotropic steel properties to the structure of a submarine's pressure hull and outer framework. The case study of the present analysis is the pressure hull of the Dolfijn submarine class. A compartment of the vessel is modelled via Finite Element Method, and its central segment is then considered. Three sets of nonlinear static analyses are performed: (i) with isotropic steel properties; (ii-iii) with orthotropic steel properties – using the Hill Potential formulation – configured in two orientations of the material. The different material formulations are shown to entail significant differences in ultimate collapse pressure values and the corresponding stresses.

NOMENCLATURE

E_L	Longitudinal direction Young's modulus (GPa)
E_T	Transversal-direction Young's modulus (GPa)
FE	Finite Element(s) (-)
L_B	Compartment's length (mm)
L_f	Frame distance (mm)
OoC	Out-of-Circularity (mm)
R_m	Mean radius of the pressure hull (mm)
P_y	Pressure causing the first yield (MPa)
P_{ult}	Pressure at the last stable step (MPa)
σ_{vm}	Von Mises stress (MPa)

1. INTRODUCTION

It is clearly acknowledged that the pressure hull constitutes one of the most important structural elements of submarines design. From the naval architect's perspective, it presents a major contribution to the displacement, and at the same time the structural analyst faces the challenge to minimize the weight and meet the safety requirements. Structural analysis in the design phase is based on specified properties of the chosen material. Following the classification Rules, for instance Det Norske Veritas (DNV) [1], the applied material may exhibit nonlinear behaviour with a proportionality limit as a percentage of the yield stress. However, the material is usually considered to be isotropic although the rolling direction may significantly affect its properties. The resulting anisotropic behaviour is recognized [2] but did not find its way to basic pressure hull design. In particular, in [2], the author studied the collapse occurrence in pressure hulls aiming to estimate the confidence interval of the collapse prediction; the residual stresses due to rolling are mentioned as key input information to consider in order to achieve a reasonable prediction. These rolling, bending, and welding processes of submarines' structural elements induce residual stresses that translate into nonlinearities in the material properties, as presented in [3]. Therein, the residual stresses are introduced in the FE model as pre-load boundary conditions. Nonlinear material and plastic behaviour of pressure hulls are studied in [4], where numerical results are compared against experimental data, on the matter of instability due to buckling.

Adopting nonlinear material properties is normally applied to the isotropic formulation, whereas anisotropy constitutes an additional level of complexity. The abovementioned phenomena (rolling, bending, welding of hull and framework plates) causes anisotropic effects, which should be accounted for given the magnitude levels of the pressure exerted on submarine hulls. Therefore, since the material properties may differ in perpendicular directions (namely, orthotropy scenario), it is important to orientate the plates in the right direction in order to align the optimal properties with the most severe stress directions. For ring-stiffened cylinders, this will correspond to the hoop (namely circumferential) direction.

The importance of orthotropy played an important role in a Nevesbu's previous project, where a significant gap hatched out between experiments and numerical simulations. A tensile test revealed the presence of a much smaller Young's modulus than the standard value for steel (181 versus 206 GPa). Furthermore, one direction showed an ideal elastic-plastic behaviour, while along the perpendicular direction the proportionality limit was reached at 75% of the yield stress. This orthotropic behaviour is applied to an actual submarine design, although the material under consideration was S350. Nowadays, pressure hulls are manufactured with higher yield materials such as HY100. However, to avoid confidentiality issues, the influence of orthotropy is demonstrated by a classic Dutch submarine design, namely the Dolfijn class [5]. This three-cylinder design was built in the 1950s, with steel having a yield stress limit of 353 MPa and Young's modulus of 206 GPa (St 52).

This paper presents the results obtained from the utilization of orthotropic steel properties in terms of ultimate pressure and the related stress occurring in the critical location within the model. The outcomes are compared with the ones obtained by using isotropic material. The compartment of the Dolfijn class submarine is used as case study in the analysis.

2. METHODOLOGY

2.1 DESCRIPTION OF THE CASE STUDY AND THE FE MODEL

The general arrangement of the Dolfijn class is presented in Figure 1. The analysis focuses on the upper cylinder between the spherical bulkheads at frame 44 and 72. With a design frame distance of 750 mm follows a compartment length of $L_B = 21000$ mm. The mean radius is $R_m = R_{in} + T_h/2 = 1549$ mm, and this gives a ratio $L_B / R_m = 13.6$, which is rather a high value in submarine design. The relevant dimensions of the pressure hull are presented in Table 1.

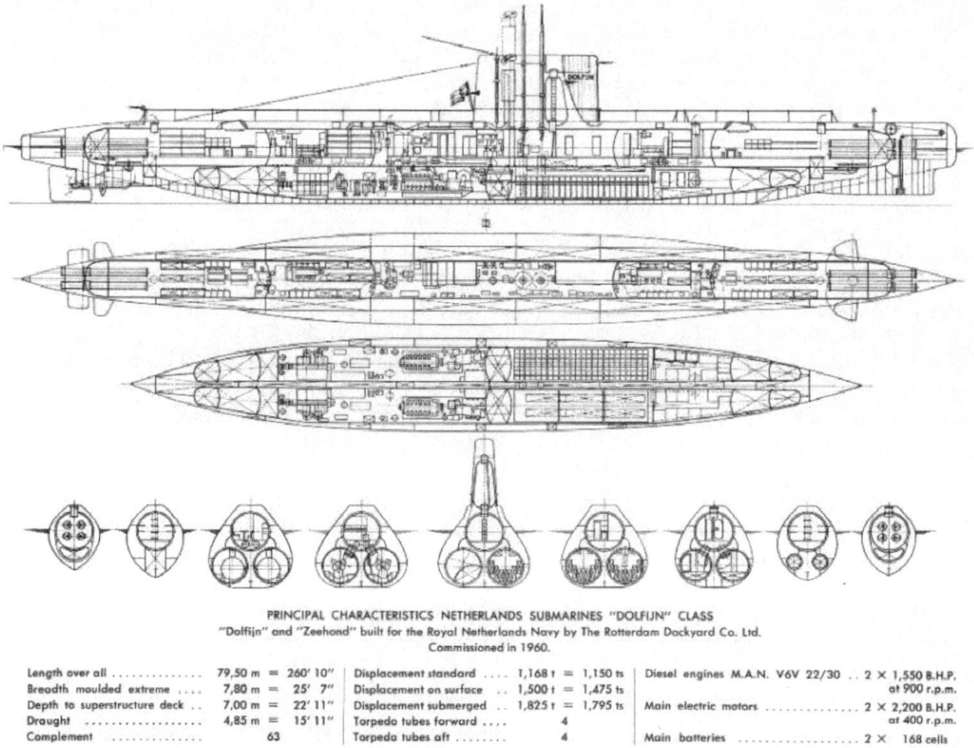

Figure 1: General arrangement of the Dolfijn class submarine.

Table 1. Dimensions of the Dolfijn class submarine.

Quantity	**Symbol**	**Value**	**Unit**
Compartment length	L_B	21000	mm
Inner radius	R_{in}	1540	mm
Hull thickness	T_h	18	mm
Actual inter-frame distance	L_f	375	mm
Web height	h_w	124.5	mm
Web thickness	s_w	10	mm
Flange width	b_f	46	mm
Yield stress (St 52)	σ_y	353	MPa
Young's modulus (St 52)	E	206	GPa
Poisson ratio	v	0.30	-

2.2 PROCEDURE

Firstly, the Finite Element (FE) model for the submarine's compartment is introduced; its global buckling pressure determines the buckling mode shape and so the number of lobes over the circumference. A reduced portion of the full FE model is then considered to be analysed in the subsequent phases of the work, in order to increase the computation efficiency while ensuring that the influence of the compartment's ends is negligible. The reduced model includes three outer frames; this system is thus employed for a nonlinear static analysis, performed by adopting the standard specifications of an equivalent steel material (St 52). In the numerical simulations, the software tool ANSYS is used for the scope. After that, orthotropic properties will be applied to the model's material, by using experimental data from a past project, in which steel specimens were tested and turned out to exhibit an anisotropic behaviour. The Hill model for anisotropic materials is adopted [6,7]. In the latter analysis, the effect of the orientation on the collapse pressure is investigated by interchanging the material directions of the two different elastic moduli and failure stress.

3. NUMERICAL ANALYSIS RESULTS

3.1 INITIAL NONLINEAR FE STATIC ANALYSIS

The compartment was modelled by means of the FE software ANSYS, and it includes spherical bulkheads at the two ends of the cylindrical structure. The T-section dimensions of the outer ring frames were determined by maintaining the same area and moment of inertia of the original double bulb sections in the Dolfijn design. In Figure 2, the undeformed geometry of the model is shown together with the contour plot of the first linear elastic buckling pressure; the deformed pattern points out two waves over the circumference. This already indicates a relatively long compartment, and with the application of nonlinear FE analysis, it is worthwhile to reduce the model due to the demanding computational effort. It can be concluded that it is justified to consider an isolated portion at the centre of the cylinder with the application of symmetry conditions. In fact, to support this operation, the normal stress in the hull sections at the border of the reduced model was compared against the stress value calculated through the simplified, static equilibrium formula for long cylinders with bulkheads at the extremities. The *reduced model*, characterized by the three frames, is shown in Figure 3, which also outlines the boundary conditions applied to the edges.

The ring frame at the centre of the compartment is positioned at frame 58, and the reduced model comprises frames 57^5, 58, and 58^5. Because of symmetry, just one half of the section is considered in the simulations. The reduced model is displayed in Figure 4 in combination with the result of the basic nonlinear analysis. The material model is ideally elastic-plastic with the properties given in Table 1. The model comprises an Out-of-Circularity (*OoC*) pattern with 2 waves over the circumference, which gives the highest bending stresses induced by the imperfection. The amplitude of the *OoC* is $0.005 \cdot R_m = 7.75$ mm. More specifically, the *OoC* is applied to the nodes of the FE model via enforced displacements, on the basis of a sinusoidal function (describing the lobes) along the angular coordinate in the cylindrical frame of reference.

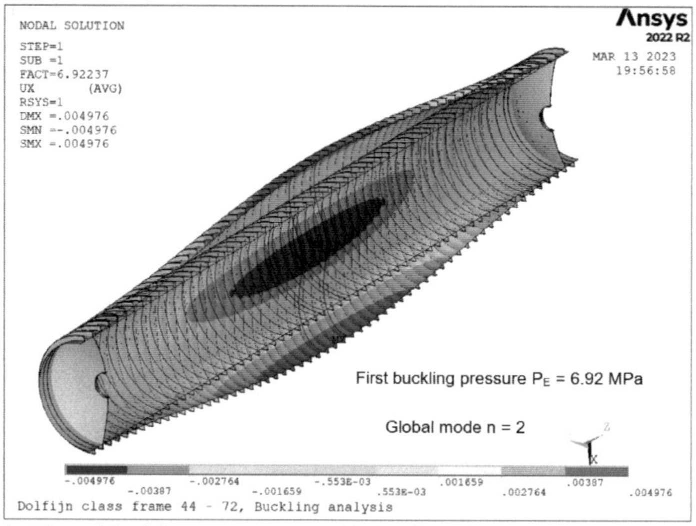

Figure 2: FE model of the compartment with the first linear elastic buckling pressure.

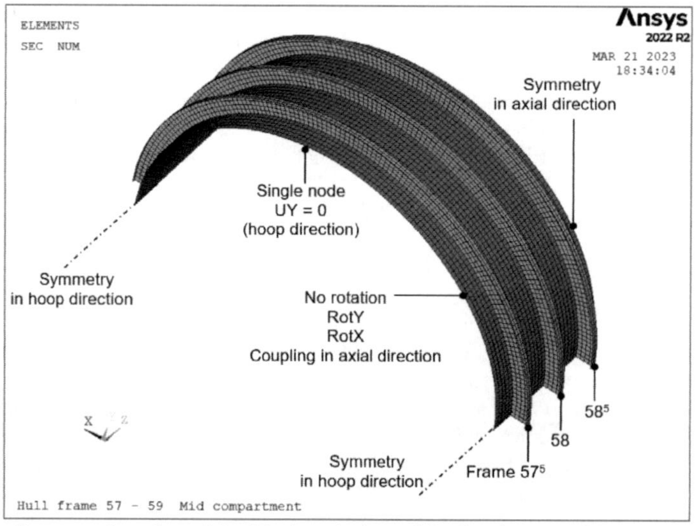

Figure 3: Three-frames reduced model with description of the implemented boundary conditions and OoC=0.005·R_m.

Figure 4(a) shows the von Mises stress (σ_{vm}) at the pressure giving first yield (P_y) at 3.345 MPa. The first yield occurs at the outer fibre of the flange and the last stable solution is found to take place at 3.505 MPa (P_{ult}). The von Mises stress versus pressure is presented in Figure 4(b), and this graph indicates little plastic reserve. The first yield occurs at 95% of the ultimate pressure, and this underpins the assumption in classification rules that global collapse is imminent when the outer fibre of the flange yields.

The design of the Dolfijn class was typically driven by global collapse scenario. Figure 5 shows the stress distribution in the hull shell, and up to the ultimate pressure the shell remains in the elastic regime. The location with the maximum stress is indicated in Figure 5(a), and the axial and hoop stress are presented in Figure 5(b). As expected, bending over the frame support induces the largest stress in axial direction. Although the axial stress exceeds the yield stress, σ_{vm} exhibits a maximum value of 329 MPa (corresponding to the 93% of the yield stress).

From this analysis follows that it is worthwhile to consider the material properties of the flange in hoop direction.

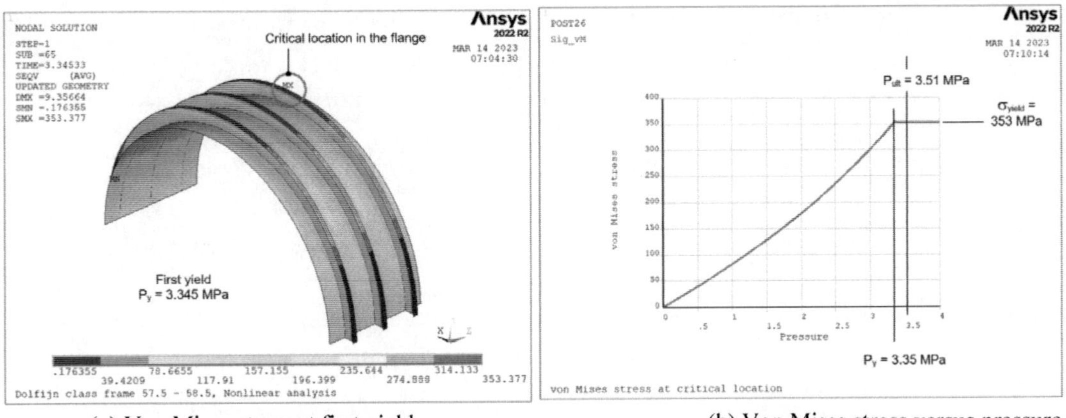

(a) Von Mises stress at first yield (b) Von Mises stress versus pressure

Figure 4: Reduced FE model with the results from the nonlinear analysis.

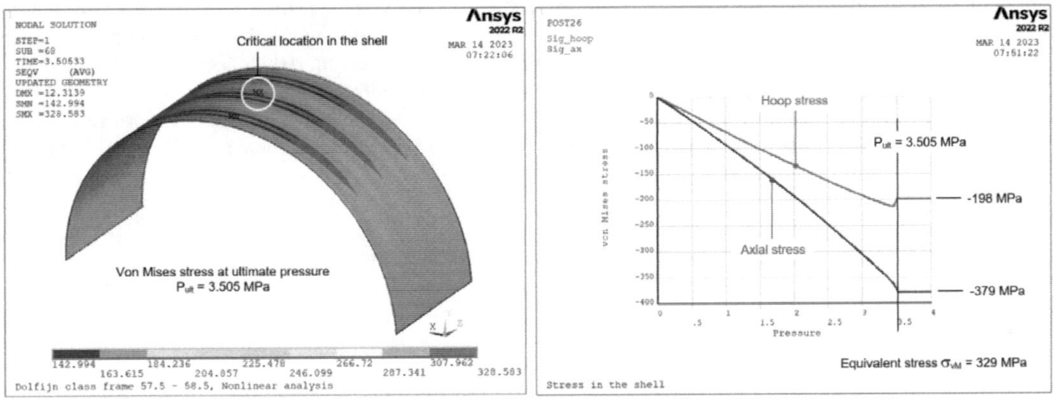

(a) Von Mises stress in the shell at first yield (b) Axial and hoop stresses versus pressure

Figure 5: Nodal von Mises stress distribution in the hull shell.

3.2 NONLINEAR ANALYSIS USING ORTHOTROPIC STEEL

In a past project developed by Nevesbu, a 5-mm-thick specimen of steel plate was analysed to determine its structural properties, revealing the presence of two different elastic moduli along perpendicular directions. Figure 6(a) shows the stress-strain curve for the longitudinal direction of the plate, resulting in a Young's modulus value of 191.4 GPa (E_L) and a yield stress limit of 448 MPa; besides, Figure 6(b) shows the curve for the transverse direction, which gives an elastic modulus of 181.0 GPa (E_T) and a yield stress limit of 410 MPa. Hence, a different formulation of the material properties and for the yield criterion are necessary in order to take into account the anisotropic behaviour of the steel. In this phase, the Hill Yield Potential model is employed [6, 7]; specifically, the Hill formulation defines a surface in the stress-state domain in which the first yield of the material occurs, and the different moduli along the principal axes determine this surface through coupling terms, which involve all three spatial directions [7]. The Hill's theoretical model for anisotropic metals demonstrated to deliver reliable results in terms of criterion to predict the plastic failure [6, 7]. In the analysis herein, a biaxial stress state is considered.

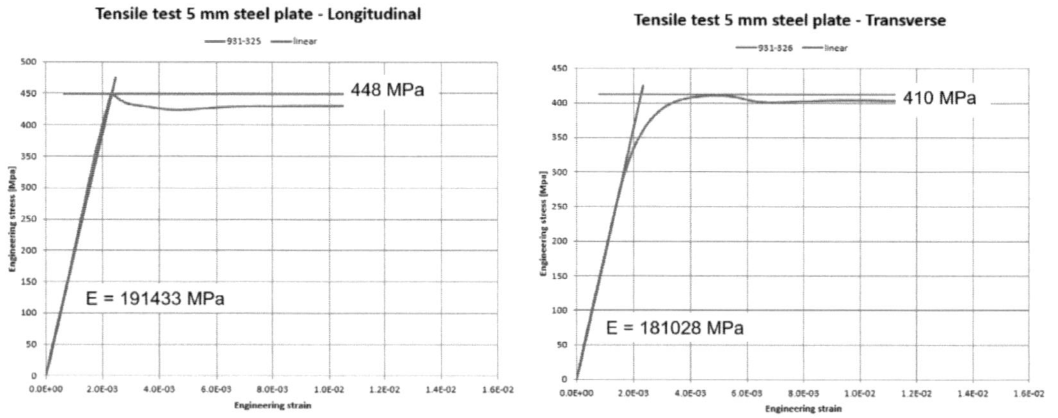

(a) Longitudinal direction test (determining E_L) (b) Transverse direction test (determining E_T)

Figure 6: Tensile tests: outcomes obtained from the specimens.

The isolated central portion of the structure is considered at this stage as well. In all shell elements constituting the hull and the outer framework, E_L and E_T were applied to the two planar directions of the shell coordinate system; the shell thickness direction is considered to not contribute to the global yield phenomenon, thus the coupling terms in the Hill potential matrix are all equal and obtained through the planar shear stress. In the bi-linear formulation of the stress-strain relations (for both E_L and E_T), the slope of the line in the plastic regime is set to be equal to $E_L/1000$, which takes effect *after* reaching the yield stress (respectively) when the linear elastic regime ends. The bi-linear modelling option is adopted for the E_T case even if the experimental curve looks nonlinear before reaching the yield limit (Fig. 6(b), blue line).

Two simulations were performed, in which the orientation of the two different elastic moduli (E_L and E_T) was rotated by 90 degrees, thus swapping the material's direction in each shell element. The boundary conditions and loading application were maintained as implemented for the isotropic steel case. Figure 7 shows the outcomes with 0-deg material orientation, i.e., with E_L in place along the circumferential direction of the hull, the frame's web, and flange. The pressure level determining the collapse results to be equal to 3.87 MPa, with a maximum equivalent σ_{vm} of 448 MPa occurring at the topside of the flange of the frame 58[5].

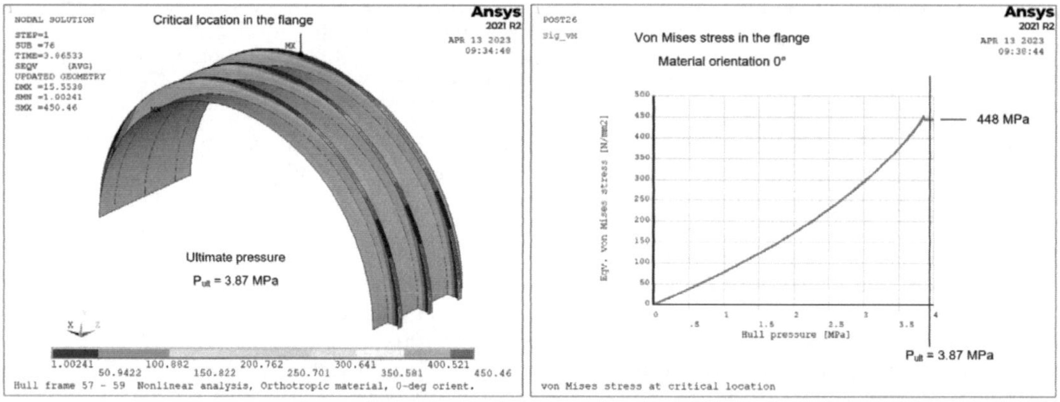

(a) Equivalent von Mises stress at P_{ult} (b) Von Mises stress vs. hull pressure

Figure 7: Reduced FE model with the results from the nonlinear analysis with orthotropic steel (0-deg orientation).

The results obtained with 90-deg orientation (E_L present along the axial direction of the hull shell, radially along the frames' webs, and axial along the frames' flanges) are shown in Figure 8. In this case, P_{ult} resulted to be 3.61 MPa and the maximum σ_{vm} reaches 410 MPa.

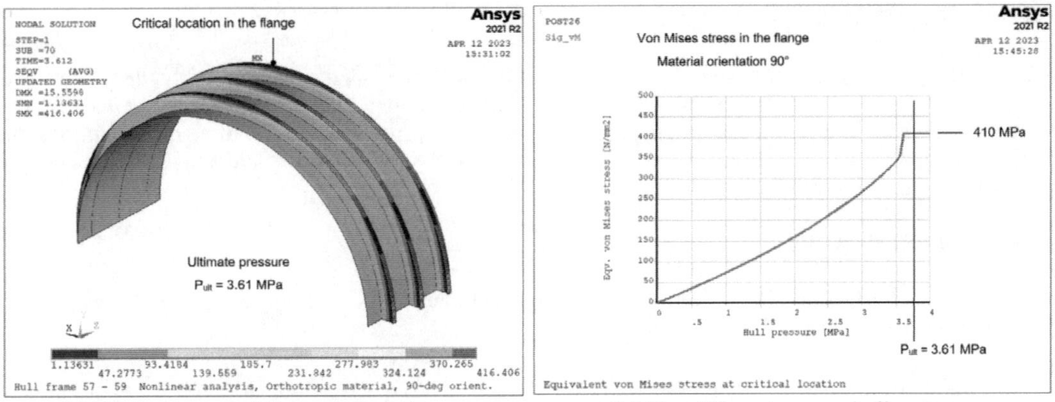

(a) Equivalent von Mises stress at P_{ult} (b) Von Mises stress vs. hull pressure

Figure 8: Reduced FE model with the results from the nonlinear analysis with orthotropic steel (90-deg orientation).

4. DISCUSSION

4.1 COMMENTS ON THE RESULTS

Table 2 reports the outcomes of the two orthotropic-material simulations, listing the ultimate pressures and the corresponding von Mises stresses occurring at the moment of the respective collapses. The mutual differences among those two quantities as the material setup changes are significant, and the orthotropic case with E_L along the hoop direction demonstrates to give the highest resistance to the external pressure. The yield stress 353 MPa for St 52 is the minimum specified yield stress. A tensile test of the material applied for the Dolfijn class would show a higher yield stress. Therefore, those results (Fig. 4) are not entirely comparable with the ones obtained through the specimen's tensile tests.

Table 2. Main results from the numerical simulations with the reduced model and orthotropic material.

Material property	P_{ult} **[MPa]**	σ_{vm} **[MPa]**
Orthotropic (0-deg orientation)	3.87	448
Orthotropic (90-deg orientation)	3.61	410

In all simulations (including the one with St 52 steel), the structure showed to fail in correspondence of the outer rings' flanges, whereas the hull shell's stress distribution remains in the elastic domain. This confirms that the design was conceived to counter global collapse with no insurgence of local, inter-frame plastic failure. Therefore, the overall structure of the hull and outer framework could potentially be optimized in order to reduce the hull's shell thickness and thereby: (i) let the local, inter-frame failure occur first; (ii) reduce the structure's weight.

4.2 FUTURE DEVELOPMENTS

Further configurations of the orthotropic material orientation could be studied. Firstly, the E_L and E_T directions may be referred to the original plate rolling direction used for the nesting process during manufacturing; to model this scenario, in each shell element of the ring frames (especially the webs) the material is oriented according to the global reference system and not along the curvature of the hull. Nonetheless, the highest strength shall be applied along the hoop direction. Secondly, the different components (hull, web frame, flange) may be assembled with different mutual orientations of the material. Multiple combinations can then be arranged, using the orthotropic properties of the steel to align hull and framework independently. In addition, the possibility of having a high Young's modulus corresponding to a lower ultimate stress could be investigated: this case should be based on experimental data on further steel specimens.

5. CONCLUSIONS

The analyses presented in this paper have pointed out that the choice of the material orientation – specifically, orthotropic steel – for submarine hull structure applications is important, as the resulting external pressure causing the material to yield changes significantly. In the design of submarines, these considerations can be taken into account to optimize the characteristics of the pressure hull's structural components in relation to the design operational conditions.

6. REFERENCES

1. 'DNV Rules for Classification – Underwater Technology', *Part 3 Pressure hull and structures*, Chapters 1 to 3, Edition July 2021.

2. GRAHAM, D., 'Predicting the collapse of externally pressurized ring-stiffened cylinders using finite element analysis', *Marine Structures,* Vol. 20, P. 202-217, 2007.

3. FRANQUETTO, P R., NETO, M M., 'Finite element cold bending residual stress evaluation on submarine pressure hull instability assessment', *Journal of Offshore Mechanics and Arctic Engineering*, Vol. 138, Issue 6, Art. N. 061101, 2016.

4. ROSS, C T F., BULL, C., AL-ENEZI, M., LITTLE, A P F., 'Geometrical and material non-linear analyses of model submarine pressure hulls', *Civil-Comp Proceedings*, Vol. 93, Code 102643, 2010.

5 Museum ships, https://www.marinemuseum.nl/nl/collectie/museumschepen/, recent access: April 21, 2023.

6. ANSYS 'Mechanical APDL Theory Reference, Par. 4.213' *Specialization for Anisotropic Plasticity (Hill Potential Theory)*, Release 18.2, August 2017.

7. HILL, R., 'A theory of the yielding and plastic flow of anisotropic metals'. *Proceedings of the Royal Society of London*, 193:281–297, 1948.

7. AUTHORS BIOGRAPHY

Jack J. Reijmers is a retired specialist engineer, still attached to Nevesbu. He graduated in 1982 at the Delft University of Technology in Naval Architecture and joined Nevesbu since then. He participated in the design of the Walrus, Sea Dragon and Moray class submarine. For several years he worked as a consultant for Navantia in Cartagena (Spain).

Alessandro Zambon, PhD works as a marine-mechanical engineer at Nevesbu. His expertise area includes ship propulsion, vibration analysis, marine structures, and numerical modelling. His previous experience regarded research on structural dynamics and FEA; he specialized on torsional vibrations along shaft lines under ice-propeller interaction at Memorial University of Newfoundland (St. John's, Canada) in partnership with the American Bureau of Shipping (ABS).

LEVERAGING CROSS-INDUSTRY APPROACHES TO ENABLE SUBMARINE DISPOSAL THROUGH IMPROVED DESIGN

K Dowrick and **J Sheasby**, BMT Ltd

SUMMARY

Decommissioned submarines yet to be disposed of, represent a huge logistical and financial burden to the UK Ministry of Defence (MoD). Overcoming the barriers that delay disposal, require the development of approaches that facilitate the simplification of recycling for the 90% of the platform that is made up of recyclable metals. This challenge is not isolated to the submarine community but seen throughout the engineering sector from automotive to the built environment. Given this common challenge, this paper considers some of the core lessons we can take from other sectors.

The element that makes the submarine disposal question slightly more unique is the presence of radioactive material and the tight space boundaries on board. This paper splits out the two elements as nuclear and non-nuclear waste, identifying ways the whole lifecycle including disposal can be incorporated into the design analysis. It will examine the potential of geomelting radioactive substances into a glass like substance and explore approaches that can be adopted to simplify recycling for decommissioned submarines, including the use of versatile substances, reducing different materials, as well as the classification of parts during design.

1. INTRODUCTION

1.1 THE ISSUE OF RADIOACTIVE AND NON-RADIOACTIVE WASTE

Due to their complex composition of materials, and unprecedented logistical challenges, full dismantling of nuclear submarines can pose significant challenges. By splitting this waste into two categories – radioactive and non-radioactive, it might be possible to unpick the issues produced by these types of waste. However, when it comes to disposing of this waste, it isn't quite as simple as splitting it into two categories and throwing it away. Everything must go somewhere, but finding the most economically viable and environmentally sustainable approach is a priority.

Aside from the nuclear waste, the remaining 90% volume of submarine materials, such as steel, are recyclable or able to be repurposed [11]. It is important to note that even with endless recycling options and high recycling rates, it is not a stand-alone solution. Although recycling is an alternative to landfill, there are also multiple limitations, which contributes to the financial burden on the MoD. Recycling sorting facilities are expensive in themselves, and it is becoming more expensive to recycle rather than buy new materials, this is due to sorting facility running costs, storage, pollution and complexity of materials and chemicals. We need to look more holistically at the part that recycling plays in the current disposal cycle and what stage it currently sits at as an option and as a political priority. One way of analysing this is looking first to the waste hierarchy pyramid, and initially where consumption can be reduced in the 'Prevention' stage.

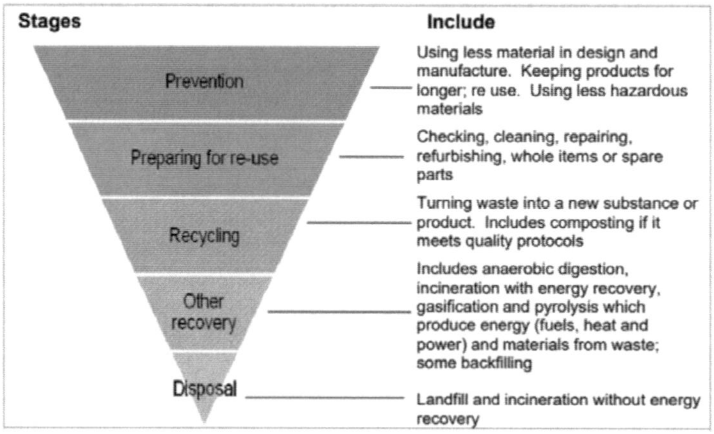

Figure 1: Waste Hierarchy Pyramid [1]

Submarine production is a necessity to protect our nation, therefore the need for them means that production will not be halting for the foreseeable future. Looking to the 'preparation for re-use' stage of the pyramid, it breaks down existing submarines to reuse their parts such as unspent nuclear fuel and metal structure. By starting with the end in mind, the design stage is where effort should be concentrated to build platforms to be dismantled. By building platforms to be dismantled, we move away from a linear economy into a more circular economy.

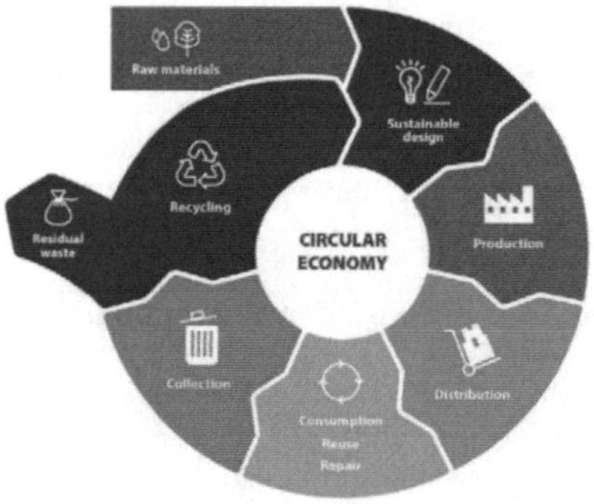

Figure 2: Circular Economy [2]

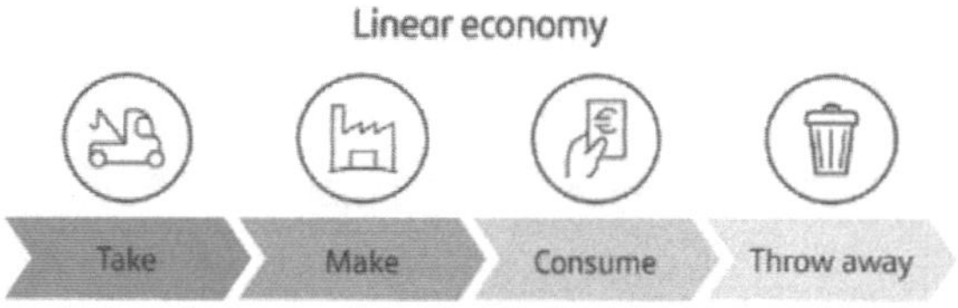

Figure 3: Linear Economy [3]

The Linear Economy is a model of the economic process that currently dominates our society with a '***Take*** (raw materials and natural resources), ***Make*** (into products for consumption), ***Consume*** (products sold to society), ***Throw*** (products are thrown away due to planned or perceived obsolescence). Planned obsolescence is the plan of an industry where the products are made to become obsolete. Apple's 'Batterygate' with the deliberate slowdown of iPhones is just one controversy of many [28]. This is either functionally, (they are made to break or slow down), or they are perceived to be obsolete. Although the product may still be functional, it is perceived to be needing to be replaced, this could be down to simply being out of style or unfashionable. The Circular Economy is a different economic model which provides an alternative to the linear economy. The circular economy implies a conception or system that keeps the added value of a product as much as possible and eliminates the waste [4]. According to the Ellen MacArthur Foundation [5], the circular economy tackles climate change and other global challenges like biodiversity loss, waste, and pollution, by decoupling economic activity from the consumption of finite resources.

1.2 WHERE DO SUBMARINES SIT WITHIN THESE MODELS?

The difference between submarines and many of our day-to-day products, is that submarines are built to last. They are designed for a service life of more than 25 years and therefore they need to be designed to meet the needs of tomorrow as well as today [6]. Despite being built to last for over 25 years (therefore sitting between the two models), how they are disposed of is not something that historically has played a huge part in the design process. It is possible to move towards a Circular Economy model by designing submarines to be taken apart, something that is often hindered by the high performance and space constrained design space. If more analysis could be provided in the design stage, the process of getting the materials to recycle could be simpler; and by carefully selecting the number and type of materials during design, the recycling process could be simplified. Many of the material recycling issues seen on a submarine are not unique and lessons can be learnt from other sectors such as automotive and construction; however, one area that does cause significant added complication is the presence of radioactive/nuclear waste. Disposal of nuclear waste is arguably more black and white than the disposal of other materials. There is spent or unspent nuclear fuel and there are the materials that have become contaminated being in the same compartment as the fuel. This is split into levels and is dealt with accordingly. Other issues associated with radioactive waste are long-term contamination, where it is being stored and pollution of the surrounding air, land and water.

1.3 WHAT IS CURRENTLY BEING DONE

The UK Government's commitment to provide a safe, cost-effective, and environmentally responsible process for dismantling and disposing of the 27 nuclear-powered submarines, is due to be delivered through the Submarine Dismantling Project (SDP). The SDP is managed by the Submarine Delivery Agency (SDA), a submarine formally enters the SDP once its nuclear fuel is removed, and the fuel is transported to Sellafield to be stored [7]. However, progress remains slow [7] and Submarine defueling was suspended in 2004, awaiting facilities and infrastructure upgrades, subsequently delaying the process of submarine dismantling.

As an example of design limitations, between 2016-2018, low-level radioactive waste was removed from *Swiftsure* class platforms, within the budget and time frame. Due to the success of this, removal of low-level radioactive waste soon began on *Resolution* class platforms. Removal of intermediate-level radioactive waste on *Swiftsure* has not begun, as the designs of removal, transport and storage processes have not been confirmed [7].

As the WARSHIP 2021 [6] paper discussed; the full disposal of a platform and onboard equipment should be acknowledged throughout the Concept, Assessment, Demonstration, Manufacture, In-Service and Disposal (CADMID) lifecycle of a platform. Consideration should be given to how the materials of a particular piece of equipment (whether being procured in the design stage or as a design change part way through a platform's life) shall be handled and sustainably disposed of at the equipment or platforms end of life. The MoD aims to comply with The Waste Hierarchy [1], with the prevention of waste being of the highest priority, but as current decommissioned submarines cannot be redesigned, their disposal shall focus on reuse and recycling of materials. The current submarines awaiting disposal were built 20 to 30 years ago, and designed years before, when environmentally safe disposal routes were not as developed as today, and acceptable levels of waste disposal were very different to what is acceptable today in accordance with legislation. Moreover, future submarine design should incorporate a standardised plan to procure materials with the lowest risk of waste when end of life occurs, or at least make this a contributing factor to the choice of material.

2. NUCLEAR WASTE

The issue with this current method of submarine disposal is even if 100% of the other non-nuclear materials get recycled, we are still left with spent nuclear waste, which requires complex storing due to the emission of heat and radiation, where it is currently contained largely at Sellafield.

A solution to the costly and environmentally damaging processes that currently comprise submarine disposal, could be geomelting – based on the principle of vitrification, converting something into glass or a glass like substance via heat. Vitrification of High Level Waste (HLW) has been studied in the UK since the 1950's, this work has been culminated in the commissioning of Waste Vitrification Plant (WVP) lines 1 and 2 at Sellafield in 1990, and line 3 in 2002 [9]. Geomelting is a step further by combining this dangerous waste with a clean material such as soil, minerals and or glass frit (about an 80/20 ratio of clean material to waste) then converting it into a glass like substance. This creates a leach proof material that is harder than concrete [10]. When the glass is molten it is combined with the nuclear waste, when it solidifies, the waste becomes part of its molecular structure, so even if the glass were to shatter, the radioactive materials would not leach out. However, it is currently still regarded as radioactive, therefore will still require some storage.

2.1 GETTING HOLD OF THE WASTE

To treat this radio-active waste, it needs to be removed from the platform. Which also needs to be separated from other 'non-nuclear waste' that the rest of the submarine consists of; this is a extensive process carried out by gas axes and cutting tools. Currently, when a submarine reaches the end of its lifecycle the nuclear waste onboard needs to be collected.

In order to remove nuclear waste the submarine needs to be towed to a secure de-fuelling dock where the primary circuit can be drained and the reactor pressure vessel head removed. Each fuel assembly is then taken out and placed in spent nuclear fuel casks and put on secure units for transportation for final treatment and storage. "*During submarine operations, nuclear reactions cause radioactive substances to be generated in the pipework and components within the reactor compartment*" [11] . Thes components which consists of but not limited to – the reactor, steam generators, pumps, valves, and piping can then be removed and processes as either Low Level or Intermediate Level radioactive waste[11].

The current method for dismantling submarines, in the UK, is in three stages and is predominantly carried out by the SDP. The three-stage dismantling method allows the less hazardous parts of a submarine's nuclear reactor core, known as Low-Level Radioactive Waste (LLW) to be removed first, followed by the reactor pressure vessel (RPV) which is classed as Intermediate Level Radioactive Waste (ILW). Once removed it is transported to Sellafield where the unspent fuel could be incorporated into cores for future submarines. Once the nuclear fuel and radioactive waste at various levels has been removed, to dismantle what remains of the submarine, the SDP follows some principles of the Circular Economy such as re-use and recycling. 90% of materials consisting mainly of steel and other metals, can be re-purposed on other in-service submarines and the remaining 10% of materials goes through conventional recycling at licensed ship-breaking sites which forms the third stage [11].

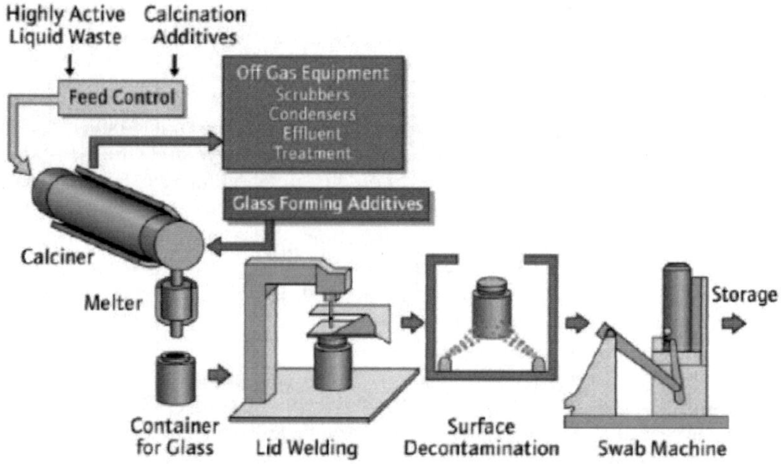

Figure 4: Schematic of the HLW vitrification process on the Sellafield site [9]

2.2 CHALLENGES

In addition to the SDP, vitrification is currently happening on a similar scale, processing nuclear waste from UK nuclear reactors at Sellafield using the PUREX method (Plutonium Uranium Extraction) [9]. Due to being quite nascent, it is not yet known if this technology exists on such a scale that it could be applied to all submarines being decommissioned. Even if this technology could keep up with the number of submarines being decommissioned, the technology that would be needed to scale up the process is unknown.

3. NON-NUCLEAR WASTE

3.1 THE CHALLENGE OF COMPOSITE MATERIALS

A common issue across many sectors is the sustainable disposal of mixed material (otherwise known as composite materials) products. For the purposes of this section we are defining composite materials as any material that is created through the combination of two or more base materials together including but not limited to all resin & fibre combinations and alloys. These products are notoriously difficult to recycle as the various materials must be isolated to be effectively identified, processed, and reused.

Although still in relatively limited quantities, since the 1960s, composite materials have been used in RN submarines, as the MoD strive to reduce the acquisition and maintenance costs and to improve operational and structural integrity of the platforms [12]. The fibres and resins used in composite materials are difficult to separate, ultimately making them difficult to recycle. Obtaining pure streams of materials that can be used to make new products is difficult and can often result in low quality recycled products or waste.

In the aerospace industry, the complex equipment and processes needed to recycle composites can be expensive, and most likely produce low quality materials that are not suitable for reuse in high-performance applications [14]. Composite materials are historically evaluated based on their performance, quality and cost, and were perceived as an 'environmentally friendly' option as they allow for lighter platforms and reduced fuel consumption. However, an increase in understanding determines that the processes of manufacturing composite materials and the recyclability challenges at the end-of-life stage outweigh these operability benefits [15]. The aviation sector aims to reduce the quantity of composite material waste by incorporating solutions into the design process, as this sector focuses on sourcing and utilising high-level and complex materials and equipment, the marine sector can follow suit.

4. SMART MANUFACTURING

The introduction of smart-manufacturing processes through advanced digital capabilities, can increase productivity and decrease costs. Smart manufacturing refers to the use of automation, analytics, and advanced technologies in manufacturing processes, such as Artificial Intelligence (AI). The development of these smart processes allows manufacturers to have 100% visibility and traceability [15]. Specifically, material waste produced in the manufacturing process can be reduced using automated cutting plans. These cutting plans become dynamic and transform in real time, based on immediate changes in production planning or the manufacturing floor. Furthermore, material rolls can be cut into varying part types simultaneously, with minimal waste, resulting in over 10% greater material utilisation [15].

5. DESIGN FOR DISASSEMBLY

Adapting the design of a piece of equipment or product based on a chain of dismantling and recycling steps at the end of life, is more effective than attempting to configure dismantling processes around a product. Economic and technical limitations mean that it is extremely difficult, if not impossible, to separate and recycle all components of a product, therefore, it makes practical sense to choose materials and assembly processes that are best suited for dismantling [23]. Incorporating design for disassembly (DfD) into concept and design phases would reduce energy consumption and carbon emissions as the requirement for raw materials would be reduced. The construction industry have trialled the DfD method and a significant challenge faced is the speed and cost of the process, as demolition is more efficient and cost-effective than dismantling a construction apart piece by piece [24]. However, research has demonstrated that dismantling could be cost-competitive with demolition if there are enough recoverable materials with good market value to offset the labour costs [25].

DfD requires an in-depth deconstruction plan including a review of the materials and components and how they should be reused or recycled. A pioneer in the field of recovered building materials, Roter Deconstruction, disassembles, conditions and sells materials [26], whereas Arup utilises the cloud-based building information models (BIM) to track materials throughout their life cycle [27].

Table 1. Required extraction of economic quantities of some materials (in g/min)

Precious Metals	**Extraction of economic quantities g/min**
Gold	0.05
Palladium	0.14
Silver	5.1
Metals	
Copper	300
Aluminium	700
Steel	50000
Plastics	
Polycarbonate (PC), Polymethyl (PM)	350
Acrylonitrile-Butadiene-Styrene (ABS)	800
Polystyrene (PS)	1000
Polyvinyl Chloride (PVC)	4000
Other	
Glass	6000

Furthermore, extensive research is required when selecting materials, for those that are of high quality (to be resilient enough for assembly and disassembly), non-toxic and have good recycling potential. Software can be used to evaluate materials, by generating Life Cycle Assessment (LCA) of specific input materials [24]. Research has estimated that only 10-20% of the costs and benefits of recycling are due to the optimisation of the recycling process, whilst 80-90% of these costs are influenced by design [17]. Material choice should also be dependent on the redemption price from recycling that material and how many grams per minute of material can be extracted from a waste stream to make the process economically viable. For example, referring to Table 1, to make steel recycling profitable, it must be possible to sort 50,000 grams of steel in a minute (or 24 Tonnes in a working day) from a waste stream, whereas only 0.05 grams of gold needs to be extracted from a waste stream in a minute to make it profitable. Moreover, it is difficult to predict the future market value of materials, especially as far into the future as 60-70 years after a submarine is designed and has reached its end of life, therefore, monitoring the market value throughout a platform's life cycle will help to manage realistic expectations of the value of materials.

Furthermore, choosing materials with physical properties that can be separated, is crucial to ensure that as much of a product, compartment, or piece of equipment can be dismantled in its entirety and the appropriate recycling process can take place. If materials are melded together in a product, meaning that reusable components are too difficult to reclaim; This product would subsequently have to be disposed of via the method of the attached substance that has the lowest common denominator, which will most likely be to landfill. This is not a problem specific to submarine, but it is a key lesson that submarine designs can learn from. To prevent recyclable materials going to landfill, future submarine designs need to choose materials that can be easily separated from each other, yet still aid in the operational capability. Taking the example of plastic, to allow separation by densiometric sorting of different polymers, the polymers should have a difference in density of at least 0.15 g/cm^3. Moreover, surface coatings are recommended to be avoided on metals, as they reduce recyclability by reducing the performance of the metals when reclaimed. Although this seems a simple solution, coatings are required to protect metals from corrosion, therefore, further research is required to weigh up the costs and benefits when considering the recyclability and operational capability of a product.

6. ALTERNATIVE MATERIALS

To decrease the reliance on raw materials and to make recycled materials a more viable option when manufacturing parts for submarines, innovative techniques and research is required. The automotive industry has done just this and continues to improve and trial new processes. When a vehicle reaches its end of life, plastic is a complex component to recycle as the methods used, such as shredding and crushing, are insufficiently selective, resulting in substantial loss. Polypropylene (PP) is a synthetic resin constructed through the polymerisation of propylene, which is melded or extruded into a variety of plastic products [19]. Currently, the automotive industry makes great use of this material, with European manufacturers using one million tons of it per year.

Although there is an increasing demand for recycled plastics in the automotive industry, over 80% of resources available are from manufacturing scraps. In 2011, The Renault Group enabled the use of End-of-Life Vehicles (ELVs) as another material source, resulting in a process called 'closed loop recycling', meaning materials are reused in the same industry. PP is a key material that is used in the closed loop recycling process, whereby car bumpers and wheel arch liners are transformed into directly reusable material for new car parts. Although due to the toxic fumed emitted when burning PP is unlikely to be used on submarines, there are key lessons can be learnt from this process and incorporated into the disposal of the current RN submarines. Instead of PP this same process could be applied again to High Density Poly Ethylene (HDPE), a material with a wide range of applications from casings to pipework.

Although lessons can be learnt from the automotive industry, we do need to take some factors into consideration when applying them for use on submarines. Using recycled materials in the construction of cars is a relatively new concept, hence durability and lifespan are not fully understood. Cars are designed to last an average of 17.5 years, but submarines are required to last at least 25 years in-service, therefore, it needs to be guaranteed that the materials used can withstand this amount of time and the pressures of a marine environment, to reduce costs and time on maintenance and repair. A crucial difference that submarines have is an enclosed atmosphere, resulting in only materials that do not produce noxious fumes being used in manufacturing. Cars and other vehicles do not have this limitation, meaning the list of materials that can be used is more extensive than that for a submarine.

7. CONCLUSION

Concepts and ideas of moving from a linear to a circular economy can be put into practise immediately if this practice is integrated early into the design process. When looking at new concepts and designs of submarines by picking versatile substances initially, the number of different types of materials is reduced, which immediately simplifies the re-use and recycling process. Despite being simplified there will still be a wide range of materials here due to load cases, environmental conditions and requirements of each material. The next stage of these versatile materials is to analyse how they interact with each other and how they fit together. This is crucial to the 'preparation of re-use' stage of the waste hierarchy pyramid as being able to take apart elements of the submarine is crucial to replacing broken elements, updating parts and finally at the end of the platforms lifecycle to separate parts made of different materials for efficient re-use or recycling.

Alongside or as an alternative to the circular economy is the geomelting of radio-active waste. There is not yet an established industry in this. However, the Nuclear Decommissioning Authority (NDA) are one of the global leaders in the civil nuclear sector. It is possible to work alongside the civil nuclear sector as well as the Submarine Delivery Agency (SDA) and with new build infrastructure such as Hinkley Point expanding all the time, alternatives to just containing nuclear waste are not that far away, geomelting nuclear waste is a real solution to just one of many challenges in sustainable submarine design and the UK defence sector has the infrastructure, personnel and ability to also become a global leader in defence nuclear decommissioning.

Utilising smart manufacturing techniques, such as AI, for cutting materials to the precise measurements that are required, can aid in reducing waste. Furthermore, designing for disposal (DfD) will combat a lot of the disposal and subsequent recycling issues experienced at the end-of-life stage.

A key factor for choosing manufacturing materials should be the redemption price from recycling the material and how many grams per minute of it can be extracted from a waste stream as this will make the process economically viable, meaning that manufacturers are more likely to choose this option as it has financial and environmental benefits. By factoring in the recycling profitability of a material, alongside its practical use, will ensure the maximum quantity of a platforms equipment and structure can be repurposed.

The current challenge of having pieces of equipment and other products that contain various materials melded together often means that the item must be disposed of via the method of the material that has the lowest common denominator, which is often landfill. This means that materials that could be recycled and reused get wasted and are not used to their optimum capability. A solution for future submarine designs is to utilise materials that can be easily separated and processed for recycling. The automotive industry has increased their use of Polypropylene (PP), a versatile substance that can be melded into a variety of products. It is easy to remove and transport as it can be baled up, shredded, and then put into a floating tank to separate into varying densities so the appropriate recycling can commence. Benefits aside, PP is unlikely to be useable inside a submarine as it is likely to give of noxious fumes in the event of a fire, making it not operationally viable.

Further research is also required to understand the operational lifespan of recycled materials and PP, as to minimise the need for repairs on in-service submarines, materials need to be durable for at least 30 years.

Overall, there are many techniques that can be utilised to dispose of the current submarines, chosen methods will be dependent on the financial and environmental cost, as well as the logistics and effectiveness of the operation. We can learn from the design of current decommissioned submarines and incorporate DfD into future designs to make disposal of parts as efficient, sustainable, and cost-effective as possible.

8. REFERENCES

1. DEFRA., 'Guidance on applying the Waste Hierarchy', *https://assets.publishing.service.gov.uk/government/uploads/system/uploads/attachment_data/file/69403/pb13530-waste-hierarchy-guidance.pdf* June, 2011

2. World Economic Forum, 'What is the circular economy, and why does it matter that it is shrinking?' *https://www.weforum.org/agenda/2022/06/what-is-the-circular-economy/* 14th June, 2022

3. Santander, 'Linear and circular economies: What are they and what's the difference? *https://www.santander.com/en/stories/linear-and-circular-economies-what-are-they-and-whats-the-difference* 15th September, 2021

4. SARIATLI, F., 'Linear Economy Versus Circular Economy: A Comparative and Analyzer Study for Optimization of Economy for Sustainability' *Visegrad Journal on Bioeconomy and Sustainable Development.* 2017, Pages 31-34, DOI: 10.1515/vjbsd-2017-0005

5. Ellen MacArthur Foundation, 'What is a Circular Economy?' *https://ellenmacarthurfoundation.org/topics/circular-economy-introduction/overview* 2023

6. HAMMOCK, L., 'Unlocking the Potential of Sustainable Submarine Design' *WARSHIP, https://www.bmt.org/media/6001/sustainability-in-submarine-design-20210507-full-paper.pdf* 2nd June, 2021

7. National Audit Office, 'Investigation into Submarine Defueling and Dismantling', *https://www.nao.org.uk/wp-content/uploads/2019/04/Investigation-into-submarine-defueling-and-dismantling.pdf* 3rd April, 2019

8. ROSSITER, K., and SMITH, C., 'Waste from Plymouth's nuclear submarines could be buried deep on Dartmoor' *Plymouth Live https://www.plymouthherald.co.uk/news/plymouth-news/nuclear-waste-submarines-plymouth-dartmoor-1849375* 1st August, 2018

9. HARRISON, M., 'Vitrification of High-Level Waste in the UK' *Elsevier. Procedia Materials Science 7. 10.1016/j.mspro.2014.10.003.* Pages 10-15, 2014

10. CHEMEUROPE., 'Geomelting' *https://www.chemeurope.com/en/encyclopedia/Geomelting.html* 2023

11. Gov.uk., 'Submarine Dismantling Project' *https://www.gov.uk/guidance/submarine-dismantling-project* 20th May, 2021

12. MOURITZ, A P., GELLERT, E., BURCHILL, P., and CHALLIS, K., 'Review of advanced composite structures for naval ships and submarines' Vol.53, Issue 1, pages 21-42 *https://www.sciencedirect.com/science/article/abs/pii/S0263822300001756* 2001

13. Antonio., 'Fiberglass: The Material of Choice for Submarine Hulls' *https://www.ussjpkennedyjr.org/fiberglass-the-material-of-choice-for-submarine-hulls/* 13th November, 2022

14. Plataine 'How to Reduce Composite Material Waste: Steps to Overcome This Industry Challenge' *https://www.plataine.com/blog/reduce-composite-material-*

waste/#:~:text=The%20fibres%20and%20resins%20used%20in%20composite%20materials,results%20in%20lower%20quality%20recycled%20products%20or%20waste 7th February, 2023

15. Plataine 'Sustainability in Aerospace Composites Manufacturing: How AI and IIOT Drive Results' *https://www.plataine.com/blog/sustainability-in-aerospace-composites-manufacturing-how-ai-and-iiot-drive-results/* 8th September, 2022

16. National Composites Centre 'UK to lead the development of the next generation of sustainable composite materials' *https://www.nccuk.com/news/uk-to-lead-the-development-of-the-next-generation-of-sustainable-composite-materials/* 2023

17. DESAI, A., and MITAL, A., 'Evaluation of disassemblability to enable design for disassembly in mass production' *International Journal of Industrial Ergonomics,* Vol 32, Issue 4 *https://www.sciencedirect.com/science/article/abs/pii/S0169814103000672?via%3Dihub* October, 2003

18. GALLONE, T., and ZENI-GUIDO, A., 'Closed-loop polypropylene, an opportunity for the automotive sector' *Value and limitations of plastics,* pages 48-53 *https://journals.openedition.org/factsreports/5225#tocto1n6* 2019

19. Britannica, The Editors of Encyclopaedia 'Polypropylene'. *Encyclopedia Britannica, https://www.britannica.com/science/polypropylene* 2017

20. MADDAH, H.A., 'Polypropylene as a Promising Plastic: A Review' *American Journal of Polymer Science,* 6(1): 1-11 *doi: 10.5923/j.ajps.20160601.01,* 2016

21. CANCELLI, U., MONTEVECCHI, G., MASINO, F., MAYER-LAIGLE, C., ROUAU, X., and ANTONELLI, A., 'Grape stalk: a first attempt to disentangle its fibres via electrostatic 9eparation' Vol. 124 https://www.sciencedirect.com/science/article/abs/pii/S0960308520305332 November, 2020

22. DIPATRIZIO, P., 'Maersk Ship Recycling Renovates Design Standards' *ShipLilly https://www.shiplilly.com/blog/maersk-recycling-venture-renews-ships-renovates-design-standards/* 5th August, 2015

23. REUTER, M., and SCHAIK, A., 'Opportunities and limits of recycling: A dynamic-model-based analysis' *MRS Bulletin* 37, pages 339-347. *https://doi.org/10.1557/mrs.2012.57* 9th April, 2012

24. CUTIERU, A., 'A Guide to Design for Disassembly' *Arch Daily, https://www.archdaily.com/943366/a-guide-to-design-for-disassembly* 10 July 2020

25. Environmental Protection Agency 'OSWER Innovation Project Success Story: DECONSTRUCTION' *Office of Solid Waste and Emergency Response. https://www.epa.gov/sites/default/files/2016-03/documents/innovation_project_success_story_deconstruct.pdf* November 2009

26. Roter Deconstruction & Consulting 'Roter Deconstruction' *https://rotordc.com/?utm_medium=website&utm_source=archdaily.com* 2023

27. Arup 'Leading Sustainable Development' *https://www.arup.com/* 2023

28. CLAYTON, J., 'Apple to pay $113m to settle iPhone 'batterygate', *BBC News, https://www.bbc.co.uk/news/technology-54996601* 19 November 2020

EXPERIMENTAL DEVELOPMENTS IN ASSESSING THE SEAKEEPING BEHAVIOUR OF AN UNDERWATER VEHICLE OPERATING ON THE SURFACE

M Courdier, Z Leong, J Duffy, Australian Maritime College, Australia, **J Binns, A Conway,** Defence Science and Technology Group, Australia

SUMMARY

Underwater Vehicles (UV) are widely used and optimised to operate deeply submerged. However, UV are also required to spend significant time at the surface in a wide range of sea states for mission critical tasks. These surfaced operations present constraints on the speed, stability and dynamic behaviour of the UV compared to submerged operations. As such, an understanding of the seakeeping behaviour of a surfaced UV is critical. However, data is limited and the knowledge of surfaced ships is generally not applicable. The UV's unique hull presents very specific resistance and motion response. Of particular interest is the pitch behaviour presenting a double peak response, unlike conventional surface vessels.

This paper presents the experimental developments needed to investigate the resistance and the heave and pitch motion response of a UV in calm water and in waves. A novel experimental test rig is presented, and several steps in its testing and commissioning are detailed. The case is made that a Single Post Seakeeping Setup (SPSS) presents superior characteristics for surfaced UV seakeeping compared to the conventional double-post setup approved by the International Towing Tank Conference (ITTC). Also, the model was tested in two outfitting configurations to quantify the influence of the Aft Control Surfaces (ACS) on the motion response behaviour. Pure heave and pitch decay tests were performed, facilitated by the SPSS novel capability. The data provided offer valuable contributions to the literature for the calculation of the hydrodynamic damping of a surfaced generic submarine (specifically the BB2) model with and without ACS. The rig provides the capability to investigate the seakeeping behaviour of a surfaced UV in a range of conditions.

NOMENCLATURE

ACS	Aft Control Surfaces
AMC	Australian Maritime College
COG	Centre Of Gravity
DOF	Degree Of Freedom
HMAS	His Majesty's Australian Ship
HMCS	His Majesty's Canadian Ship
ITTC	International Towing Tank Conference
LFB	Large Force Balance
UV	Underwater Vehicle
SPSS	Single Post Seakeeping System

1. INTRODUCTION

The applications of underwater vehicles (UV) range from military operations to the collection of scientific data or industrial surveys. [1], [2] These applications are typically performed at depth; hence the design of UVs is focused on the optimisation for fully submerged operations. [3], [4] Nevertheless, UV are required to navigate on the surface at some point in their mission. The reasons why include communicating, (re)orienting themselves, or recharging batteries or air supplies, recovery and deployment, getting in and out of a harbour, or even interacting with divers and surface teams for example as pictured in Figure 1. [5]

Figure 1: UV operating in surface conditions. (left) AUV autonomous recovery manoeuvre [6], (right) Special operation diver entering a submarine through its torpedo launching tube (courtesy of Ewan Lebourdais).

The free surface effects can result in significant changes in the hydrodynamic forces acting on a UV when compared to the deeply submerged condition. At the surface, a UV is subject to waves and is exposed to the same environment as conventional surface vessels. However, most of the seakeeping and resistance knowledge applicable to conventional surface vessels is not relevant to UVs due to the significant differences in their hull forms. [7] As such the study of the seakeeping performance of a surfaced UV becomes critical as it defines the ability of the UV (and the safety of its crew) to carry out its mission. [8]

The seakeeping of a surfaced UV must be studied in regard to several factors. One of them concerns the design of a UV for safety [9]. On several occasions crews have been in difficulties due to the UV surfaced motions, for example when 5 crew members were thrown overboard in March 2007 while clearing fishing line from the *HMAS Farncomb* propeller due to large unexpected motions in response to the sea state. Several other crew members were injured while the submarine remained surfaced without forward speed. [10] Another example is the accident of the *HMCS Chicoutimi* in October 2004. The submarine was sailing at the surface in heavy seas, and hatches on the bridge fin lookout chamber were open for watchkeeping purposes. Around 2 tonnes of water flooded the vessel and drowned the main power connection as a result of the high waves and motions of the platform. The first experimental program on surfaced UV seakeeping [11] followed this accident with limited studies published in the public domain since.

A second factor justifying the study of surfaced UV seakeeping is human performance onboard. Generally, submarines are able to travel on the surface in sea states corresponding to wave heights up to four metres. However, it has been shown by a recent study conducted at the Australian Maritime College (AMC) [12] that the platform motion response to waves causes a surfaced UV to experience large accelerations. The maximum accelerations were compared to two industrial standards [13], [14], which confirmed they are a source of concern for human performance onboard. The sea sickness index was used as a reference and indicated that up to 42% of a population can experience motion sickness in less than 30 minutes in the most extreme but realistic case condition studied. This case study was performed at a wave height of 3.0 m and demonstrated that in these conditions the crew becomes the limiting factor as the UV can operate in higher sea states.

Finally, it was demonstrated in exercises at sea that life rafts mounted in the casing of a submarine would be difficult or impossible to operate in almost all but the calmest conditions. [15], [16] Such life raft generally consists of a habitable compartment with access hatches in the hull, that can detach from the UV casing and float upright on its own.

This paper presents a novel experimental programme undertaken at the AMC as part of an investigation of the seakeeping behaviour of a surfaced UV. The experimental campaign was conducted in the AMC towing tank in 2022 and benefitted from the development of a new Single Post Seakeeping System (SPSS) with enhanced capabilities. The rig design, development, test and an example of its capabilities are presented here. The findings show that the new SPSS has benefits over an existing double post system commonly used at AMC and approved by the ITTC.

2. EXPERIMENTAL PROGRAM

2.1 EXPERIMENT OVERVIEW AND OBJECTIVES

A series of model-scale towing tank experiments were conducted to investigate the seakeeping behaviour of a surfaced submarine. Of particular interest were the motion response and added resistance of the submarine to a range of wave heights and wave encounter frequencies. Additionally, the phase relationship between the aforementioned variables was of interest. The data collected from these experiments were also intended for use in validating numerical modelling such as panel code and viscous-based Computational Fluid Dynamics simulations.

The measurements included the resistance force, heave motion and pitch motion of the submarine, and the incoming wave profile relative to the nose datum of the submarine. The experiments were conducted for a range of speeds and wave conditions, including variations in frequency, height and slope, in a straight-ahead heading at a single load condition.

2.2 TEST FACILITY

The tests were conducted in the AMC Towing Tank. The tank measures 3.55 m wide, has a nominal water depth of 1.50 m and is 100 m long. The tank is filled with filtered fresh water, with the temperature recorded on every run and found to be 16.5 °C ±0.3. The tank is equipped with a flap-type wavemaker on one end and a wave absorber on the other end. The motorised carriage is fitted with data acquisition and processing systems. The speed of the tow carriage was measured using continuous recording from a rotary encoder on one of the wheels. The carriage has a maximum speed of 4.6 $m.s^{-1}$, and the constantly measured speed was within the recommended tolerances as set by the ITTC. [17]

2.3 BB2 GENERIC SUBMARINE MODEL

The submarine geometry used for the experimental campaign is the BB2 [16], [18], [19] pictured in Figure 2. It is a generic, conventionally (diesel) powered submarine geometry widely used for research purposes. The dataset generated with the submarine model can thus be shared with other research programs and compared with other data sources. The choice to use the BB2 over another UV geometry was based on the highly realistic and appropriate hull shape of the model which was provided in kind by the Defence Science and Technology Group (DST of the Australian government. Additionally, there were no other comparable models that met such high precision requirements for the experiments.

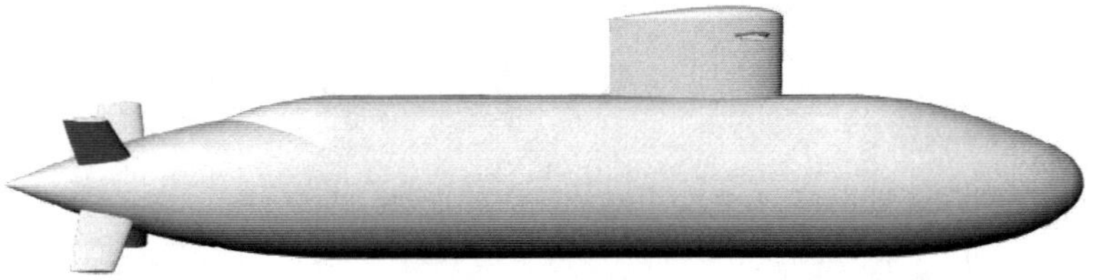

Figure 2: BB2 submarine geometry [20]. Note that the propeller is excluded from the current study.

The BB2 was scaled at 1:41.5 with the propeller excluded for the seakeeping experiment. This scale was chosen to minimize hydrodynamics blockage effects in the AMC's towing tank based on guidelines by the ITTC. [17] The model was designed and manufactured at the AMC for use in semi-captive experiments in accordance with the model tolerances specified in the ITTC recommended procedures. [21]

The model was initially designed in 2019 for a conventional double-post seakeeping experimental system as shown in Figure 3. The model required alteration by the addition of an "aft sail" to accommodate the aft post. This model was then modified for the current research to be used with a Single Post Seakeeping System (SPSS) and the "aft sail" removed. This modification was essential for more realistic scenarios in higher sea state conditions where the aft part of the casing was wet (up to 3.0 m). It also simplified the potential sources of error and allowed better control of the model mass distribution while still maintaining the stiffness of the rig.

The model was fitted with Hama strips for turbulence stimulation [22], and also removable Aft Control Surfaces (ACS). The model was designed to have the same hydrostatic characteristics with and without ACS [12], and this allowed the investigation of the ACS influence on the seakeeping of a surfaced UV.

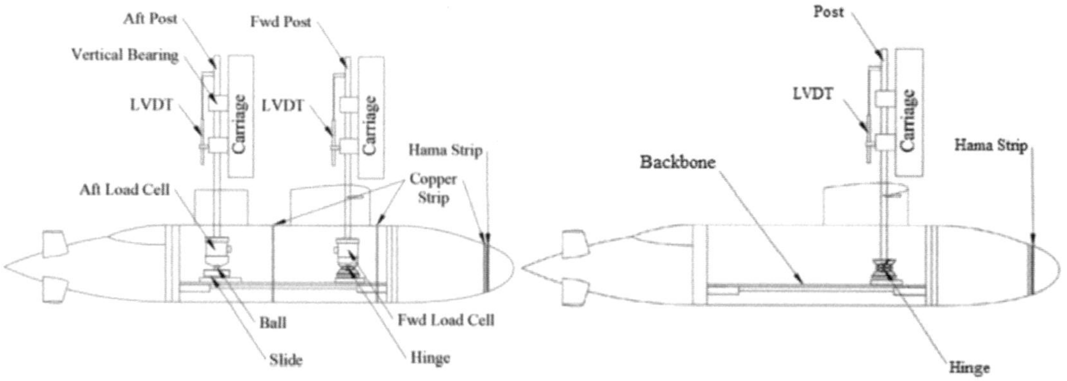

Figure 3: Experimental model setup (left) as used in 2019 with a double-post system conform to the ITTC guidelines for conventional surface vessels. (right) The Single Post Seakeeping System (SPSS) as used in the current research.

2.4 EXPERIMENTAL INFRASTRUCTURE

2.4 (a) A need for an improved test system

Given the hull shape and water plane location of a surfaced UV, there is often water running on the casing and aft of the sail when moving in waves or even in calm water at high speed. The conventional double post seakeeping system required the addition of an "aft sail" on the hull at the aft of the casing, see Figure 4. For several conditions of sea states or travel speed the "aft sail" was getting wet, hence a source of unrealistic forces and resultant motions measurement. As a result, the test conditions had to be limited, making comparison with real world applications tedious. More importantly the specific seakeeping behaviour of a surfaced UV could not be explained, as the complex hydrodynamics phenomena could not be investigated fully. Additionally, the conventional double post seakeeping system was previously developed for tests of conventional surface vessels. However, its natural frequency was extremely close to the surge natural frequency response of the surfaced UV in waves. It was thus difficult to correctly capture the resistance, also resulting in inappropriate peak motion measurements.

Figure 4: Maximum condition tested with the conventional double post seakeeping setup in 2.0 m regular head waves at 10.0 kt (full scale) due to the aft sail. Some wave frequencies corresponding to the peak motions were not achievable at this wave height. [23]

2.4 (b) Single Post Seakeeping System (SPSS)

To address the limitation of the double post seakeeping setup, a novel setup was developed for the present research. It consisted of a Single Post Seakeeping Setup with a single point attachment in the hull. It allowed the measurement of forces and motions on the UV in 3 Degrees Of Freedom (DOF): Surge and Heave translations, and Pitch rotation. The 3 other DOF were fixed through the SPSS. This novel setup allowed for the test of realistic conditions at speeds up to 14 kt (full scale) and regular wave heights up to 3.0 m (full scale). It also ensured that the peak motion frequencies were captured free of interference from the test equipment, see Figure 5. The number of instrumentation wires visible with the SPSS (Figure 5) has also been greatly reduced compared to the double post setup (Figure 4), hence reducing the risk of cable interacting with the structure or the model in motion, and greatly improving the system reliability. The SPSS went through several design iterations until determined to be suitable.

Figure 5: Newly developed SPSS in 2.0 m regular head waves at 10.0 kt (full scale) but at a peak excitation wave frequency (0.092 Hz). The aft section of the casing is clearly fully submerged at the trough of the motion.

The SPSS was attached to the model backbone through a hinge block located under the sail as shown in Figure 3. The SPSS was mounted on the towing tank carriage with the centre of the connection post in line with the longitudinal centreline of the towing tank. The SPSS was used to easily align the model at a static incident angle of 0.0 degrees (straight ahead condition). The model was free to pitch around the hinge block axis, and free to heave via the vertical post rollers. A CAD model illustrating the SPSS in its entirety is shown in Figure 6.

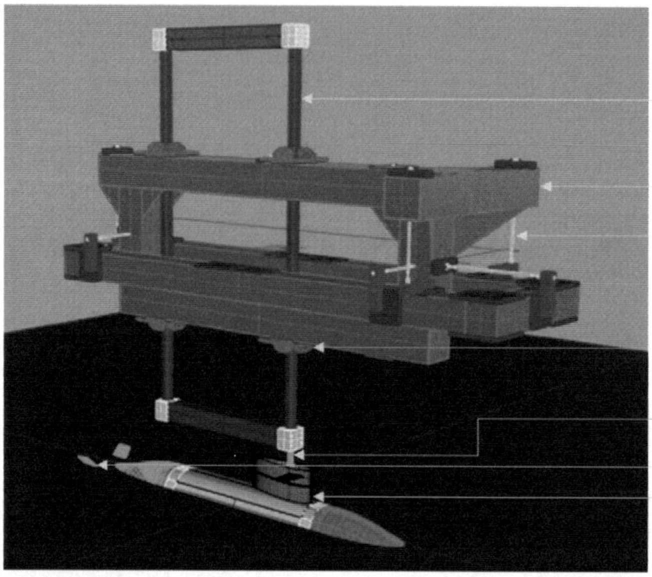

Figure 6: BB2 model attached to the SPSS fitted through its sail, and into the Large Force Balance (LFB). The LFB (depicted in red) is supported by the towing tank carriage primary structure (not depicted).

2.4 (c) Instrumentation

The force transducer used was a 6-axis Large Force Balance (LFB) developed at the AMC [24], calibrated with a 6x6 linear matrix. An incremental encoder at the hinge block was used to measure the pitch angle and was selected for its high precision and modest weight and size. A Linear Variable Differential Transformer was fitted to the forward post to record the vertical heave motions experienced by the model. The water surface elevation was measured adjacent to the UV bow using an ultrasonic wave probe. Two additional capacitance type wave probes were located at both ends of the towing tank to measure the wave quality generated at the wavemaker and potential wave reflection from the beach at the other end. A set of cameras was located around the model to visually capture the incoming waves and wakes coming off the model at a high rate (up to 200 frames per second). The data acquisition process was controlled by an in-house software added to the *LabVIEW* commercial suite and triggered automatically for recording of all sensors for calibration and test purposes. Data processing was performed through *MATLAB*, and the measurement uncertainty analysis presented in Table 1 for a test condition of 10.0 kt was deemed satisfactory. The temporal drift of the sensors was examined and found negligible.

Table 1. Measurement uncertainty summary for a confidence level of 95%

Source	**Uncertainty (%)**	**Uncertainty range**
Pitch angle	0.65	0.05 deg
Heave displacement	0.42	0.50 mm
Wave height	1.74	0.84 mm
Wave frequency	0.35	2.0E-3 Hz
Wave encounter frequency	0.42	0.02 rad.s^{-1}

2.5 TESTING PROGRAMME

The experimental programme was composed of four discrete phases. Each of them presented an incremental step to test, commission or validate the use of every component of the testing system. As such:

- Phase 1 consisted of a stress test of the novel SPSS. The test was relative to a load capability check in the form of a deflection test, and a vibration response analysis of the single post system on a dry bench.
- Phase 2 was the commissioning of the SPSS from a system point of view. It was related to the use of the SPSS with the LFB on the towing tank carriage and was designed as a risk assessment exercise. It consisted of a limited number of runs repeating conditions covered in a prior test with the conventional double post setup.

- Phase 3 consisted of a refinement of the model weight distribution, an estimation of the model's centre of gravity vertical position, as well as the measure of the model's radius of gyration.
- The last phase was the actual experimental campaign, covering 548 runs over 47 test conditions. It included some standard repeat runs for uncertainty and temporal drift quantification as per ITTC guidelines. [25]

3. SINGLE POST SEAKEEPING SETUP DEVELOPMENT, TEST AND COMMISSIONING

3.1 DESIGN RATIONALE

The aim of developing the experimental SPSS was to address the aft sail issue of the double-post system and provide an effective tool for surfaced UV experiments. It was also a design aim that the SPSS would be able to be used in the future for tests other than those relating to surfaced UV exclusively. As an added capability the SPSS can be used with different attachments to the hull. Notably, a carbon and a stainless-steel lower post were manufactured, presenting different stiffness and weight options.

The design of the SPSS began with a number of concept designs. The concept designs were mapped out in 3D CAD software providing a rapid visual check of the ideas. This also allowed for first estimations of the structure weight and guided further design options with specific materials. From these initial observations, it became clear that the success of such a single post setup would be highly dependent on the strategies for an effective strong retainment in yaw. To increase the range of operability of the SPSS the design process also aimed to provide a solution for variable angles of yaw, hence offering a capability to investigate a surfaced UV at drift angles.

From the start of the process, the SPPS was chosen to be used with AMC's LFB as a force transducer, and the dimensions of the LFB thus influenced the design steps of the SPSS. For this project, the remaining design constraints were:

- the overall SPSS mass, as the model must be floating at the correct draft (load condition) and supports the system;
- the mass distribution, as it was clear that the SPSS would fit in the hull through its sail. The sail being forward of the LCG resulted in a need to keep appropriate weights margins to fit ballast at the aft of the model for a balanced trim and pitch radius of gyration;
- the post stiffness and resistance, as by design the SPSS is the only element linking the model to the measurement devices. Of particular concern was the yaw stiffness at speed. Additionally, during the experiment, the carriage is usually backed up between runs with the model in a reverse direction even amplifying the lever arm generating sway moment;
- the SPSS natural frequency is important for surge measurement (and also a consideration for sway and yaw forces) and had to be as far as possible from the expected wave encounter peak excitation frequencies;
- the SPSS attachment point location in the hull, as it had to be located at the height of the thrust line for a realistic transfer of forces. It was clear that the attachment point would be longitudinally located forward of the centre of gravity of the model as the post was entering through its sail. That was not deemed to be an issue, and the postprocessing of the results would transfer the measurements from the actual location to the LCG if needed;
- the pitch envelope as the surfaced UV was expected to reach pitch angles in excess of ±15 degrees and the risk of possible collisions between the post and the model had to be mitigated.

3.2 FINITE ELEMENT ANALYSIS (FEA)

The main focus during the analysis was to assess the potential overall weight and stiffness of the structure, as well as its vibrating modes and response to loads exerted during testing. The SPSS was modelled with the post in three different extension conditions, and a load was applied on the hull attachment point. The extensions simulated were equivalent to when the UV was in calm water, and two extreme cases of minimum and maximum extension to cover the largest possible range of heave motion. The loads used were the mass of the model first, and a mass equal to two times the model mass to represent the added mass effects.

The result of the static analysis provides information on the deformation of the SPSS for a range of materials, and the modal analysis informed on the SPSS natural excitation frequency. The primary mode of vibration was estimated at 5.84 Hz for the carbon / stainless-steel assembly chosen for this research.

3.3 NATURAL FREQUENCY TEST

To determine the natural frequency of the designed rig in experimental conditions, a ping test was performed. The ping test is a method used to determine the modes of vibration of a structure. The test involved the excitement of the SPSS mounted on the LFB positioned on the carriage and with the BB2 model attached. The ping test was performed with four sequential hammer hits on the structure, in four different locations and directions, and the LFB force transducers and an accelerometer were used to record the effect of each impact. The output from these measurements were processed through the use of Fast Fourier Transforms (FFT) to subsequently obtain the Frequency Response Function (FRF) of the structure. The modal frequencies were easily identified from there.

All results analysed presented two resonant frequencies of the system at around 0.83 and 5.78 Hz in a transverse direction with the latter in good agreement with the frequency predicted by the FEA at 5.84 Hz, thus denoting a good agreement between both methods used. The frequency recorded in a longitudinal direction (surge force direction) was found to be very high, at 8.45 Hz.

3.4 DEFLECTION TESTS

The last test performed was a deflection test of the SPSS, intended to measure the displacement of the model in response to exerted loads. It was a simple experiment that only required the application on the model of a known load in a chosen direction and the measurement of the subsequent deflection of the SPSS. This test was performed in two configurations: a 'sway-type' deflection, and a 'yaw-type' deflection illustrated by Figure 7. For both tests, the model was constrained in pitch and heave. A comparison of the deflections of the conventional double-post setup and the SPSS was made.

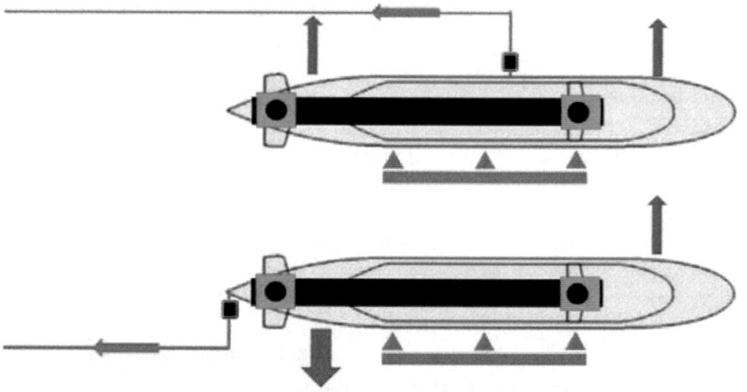

Figure 7: Top-view schematic of both deflection tests showing the tensioned line (red) with the load cell (black) and the 3 dial gauge indicators (grey arrows). The blue arrows indicate the direction of displacement of the hull. (top) The 'sway-type' force is applied at the mid body. (bottom): The 'yaw-type' force is applied at the tip of the tail.

The load was applied through a metallic wire with one end attached to a load cell connected to a calibration hole in the model's hull. While the load was applied a slight deflection of the posts occurred, resulting in a displacement of the model's hull. The displacement was measured by a set of three dial gauge indicators fitted alongside the hull. The middle dial gauge indicator was located at the LCG and two additional gauges at the same distance on each side. This allowed for a measurement of displacement in sway and yaw.

The maximum deflection to load ratio recorded for both rigs and for both tests was calculated. The results indicated that the SPSS had a higher stiffness than the double post setup in yaw (77.0%) and sway (73.5%). This validated its use for the current experimental program and was expected to result in better forces and motion capture than previously possible.

4. EXAMPLES OF RESULTS

4.2 EXPERIMENTAL DRAG MEASUREMENT

A key artefact in the measurement quality with the conventional double post seakeeping system was the natural frequency being close to the range of wave excitation frequency tested for a surfaced UV. This resulted in a noisy capture of the surge force when the double post was used as illustrated in Figure 8 (left). The visible spikes between a trough and a peak especially were considered to be a result of the excitation of the test rig resulting in an oscillating longitudinal movement. Filtering had to be used with limited success. A desirable improvement with the development of the single post seakeeping setup was to allow a higher quality surge force capture. Figure 8 (right) illustrates what was achieved on a benchmark run. The results are deemed satisfactory and presented a clear improvement.

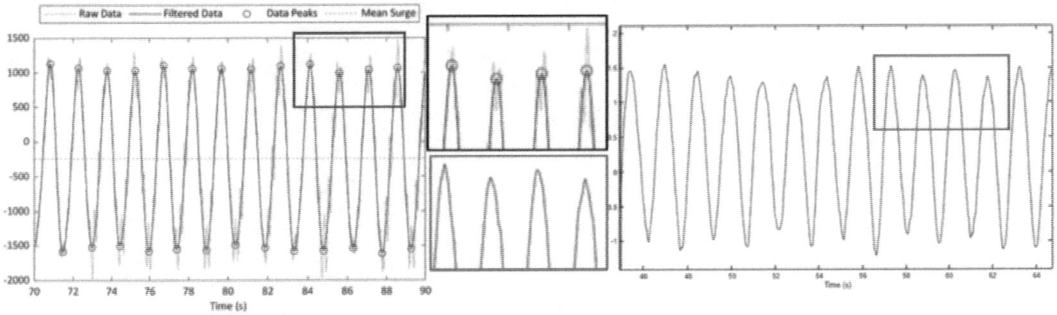

Figure 8: Example of the drag forces measurements for the (left) conventional double-post system and the (right) SPSS.

4.3 EXPERIMENTAL MOTIONS' MEASUREMENT

A comparison between the two rigs was performed on the heave and pitch motions measurements. An example is provided in Figure 9. The figure shows a shift in peak magnitude and a shift in peak frequency for both heave and pitch. For the heave motion, the peak frequency shift was 0.64%, and the magnitude shift was 1.88%. For the pitch motion, the first pitch peak frequency shift was 2.11% and the magnitude shift was 0.59%.

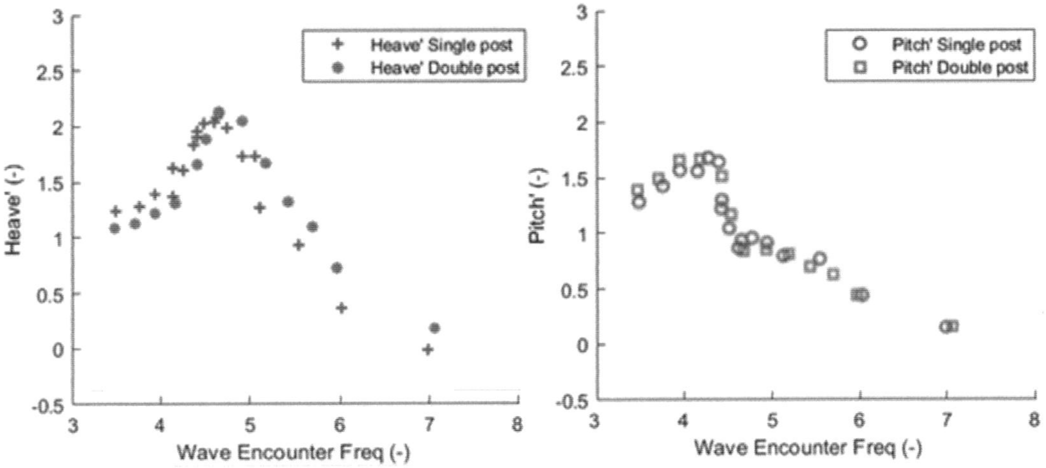

Figure 9: Non-dimensional heave and pitch motion response as a function of the non-dimensional wave encounter frequency comparison for the conventional single post and the SPSS at a travel speed of 10.0 kt in 2.0 m regular waves. Heave and pitch were non-dimensionalised based on the wave height and slope respectively.

The differences between both systems were deemed acceptable, especially given the difference in wave parameters recorded (wave frequency difference of 1.80% and wave height difference of 0.42%). The results of the SPSS are also expected to be more representative as the setup was stiffer with a higher confidence in the weight distribution of the model with the omission of the aft post. The results reported in this section are provided for clarity only but do not form the basis of an absolute quantification of the differences.

4.4 DECAY TESTS OF A SURFACED UNDERWATER VEHICLE

A design feature of the SPSS was the possibility to block each DOF independently. As such the experimental setup allowed to test the BB2 model in a free-decay test in a pure 1 DOF. This provides valuable data on the natural response of the oscillatory system for a BB2 1:41.5 scale model with and without ACS. The data can be used in panel code software for example as demonstrated in [12] or to assess the damping matrices in heave and pitch. The model characteristics were presented in section 2, the model draft was 0.198 m, the displacement was 51.710 kg, the vertical position of the COG was 0.109 m above the keel and the yaw radius of gyration was 0.404 m estimated equal to the pitch radius of gyration.

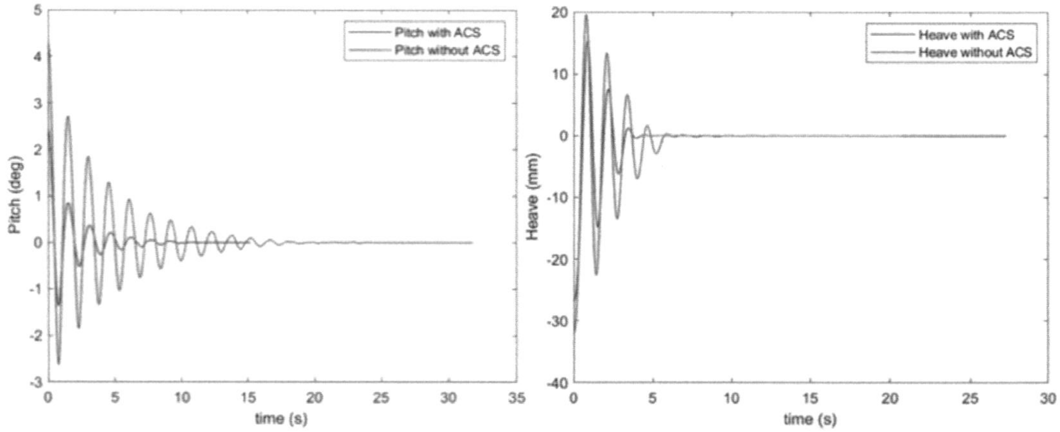

Figure 10: Impulse response recorded for the (left) pitch decay test and for the (right) heave decay test for both hulls with and without ACS.

The decay test was performed by pushing the BB2 model out of its natural resting state and releasing it suddenly. The model was pushed by a small distance only to limit the change in hull volume. The motions were recorded during the whole process and for a time allowing the system to stabilize. The heave decay test was performed by pushing down the SPSS vertically (increase in model's draft) and the initial displacement was 26.70 mm. The pitch decay test was performed by pushing down the aft section of the hull to create an initial trim angle of 2.43 deg (bow up). A FFT check was performed on the decay tests and displayed a single tone frequency confirming a pure single DOF test. The impulse responses for both tests and both hull configurations are presented in Figure 10.

5. CONCLUSIONS

As explained in this paper the conventional double post seakeeping system, although validated for use by the ITTC on conventional surface vessels was not optimal in the case of a surfaced UV seakeeping campaign. The present work thus aimed to develop a Single Post Seakeeping System (SPSS) that would address the issues identified.

The primary goal was to be able to test the BB2 model in its theoretical shape (with a single sail) and higher realistic sea states as presented in Figure 11. The SPSS allowed this while also being able to withstand the expected loads and providing a model at an optimum ballasting state. Additionally, the natural frequency of the test rig has been shifted and laid further away from the expected wave excitation frequencies to be tested, improving the quality of the forces and motions measurements. The drag force and the motions measurements were confirmed to be of adequate quality and within appropriate uncertainty levels.

It thus can be concluded that the SPSS brought a tangible improvement for surfaced UV hydrodynamics research. Future work should focus on investigating the seakeeping behaviour of a surfaced UV. Of particular interest is the double peak pitch behaviour and the negative added resistance found in waves. Significant advancement of the public knowledge are expected regarding the influence of the aft appendages on the motion response and the resistance of UV. As such, the SPSS is a valuable tool for improved future design and operations of UV.

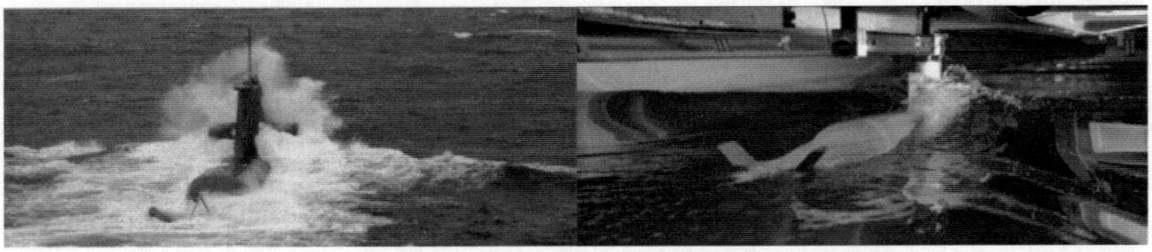

Figure 11: Realistic seakeeping conditions. (top) Collins class *HMAS Farncomb* performing pre-dive checks in 2013 [26]. (bottom) BB2 submarine at Froude number 0.196 in 3.0 m head seas during the experiments of the present research.

6. ACKNOWLEDGEMENTS

The authors acknowledge the support of the Defence Science and Technology Group of the Australian government for the use of their model during the experimental campaign. The authors also acknowledge the Institute for Marine Engineering Science and Technology for their funding towards this project through the Laurie Prandolini research fellowship. The authors also thank Mr Lewis Downie for his support to this research during his BEng final project. Finally, the authors recognise the efforts of the AMC's towing tank crew during the test, commissioning and experimental campaigns.

7. REFERENCES

1. RENILSON, M., 'Submarine Hydrodynamics'. *Cham, SWITZERLAND: Springer International Publishing AG*, 2018.

2. ALVAREZ, A., BERTRAM, 4., and GUALDESI, L., "Hull hydrodynamic optimization of autonomous underwater vehicles operating at snorkeling depth," *Ocean Eng.*, vol. 36, no. 1, pp. 105–112, Jan. 2009.

3. BOVIS, A., 'Hydrodynamique navale, 2nd Ed'. in *Coll. Les Cours. Palaiseau: Les Presses de l'ENSTA*, 2017.

4. GATES, P J., and LYNN, N M., 'Ships, Submarines and the Sea'. *Brassey's Inc*, 1990. Accessed: Feb. 02, 2022.

5. COURDIER, M., "Stanley Gray Lecture: Seakeeping Behaviour of a Surfaced Underwater Vehicle," presented at the *Stanley Gray Lecture, Sydney NSW* & online, Apr. 14, 2021. [Online]. Available: https://www.youtube.com/watch?v=K_XOAYmJjxQ&ab_channel=TheInstituteofMarineEngineering%2CScienceandTechnology%28IMarEST%29

6. KRAKEN ROBOTICS, "Launch and Recovery," *Kraken Robotics*. https://krakenrobotics.com/products/launch-recovery/ (accessed Apr. 07, 2023).

7. COURDIER, M., "Predicting wave run-up on a surfacd underwater-vehicle," presented at the *The 9th Conference on Computational Methods in Marine Engineering (Marine 2021)*, Jan. 31, 2022. [Online]. Available: http://journals.ed.ac.uk/MARINE2021

8. BHATTACHARAYA, R., 'Dynamics of Marine Vehicles'. *New York: John Wiley & Sons Inc*, 1978.

9. BURCHER, R., and RYDILL, L., 'Concepts in Submarine Design'. in *Cambride Ocean Technology series; 2. Cambridge: Cambridge University Press*, 1994.

10. LAMPATHAKIS, P., "Spy sub had China target," *PerthNow, Jun.* 16, 2012. Accessed: Mar. 09, 2023. [Online]. Available: https://www.perthnow.com.au/news/wa/spy-sub-had-china-target-ng-29ad57321665

11. HERMANSKI, G., and KIM, S., "Surface seakeeping experiments with a model of a submarine," presented at the *11th International Symposium on Practical Design of Ships and Other Floating Structures: 2010 PRADS 11th triennial, Rio de Janeiro, Brazil, Rio de Janeiro, Brazil*, Sep. 2010, p. 9.

12. COURDIER, M., J E., "Seakeeping Behaviour of a Surfaced Underwater Vehicle," *PhD Thesis, Australian Maritime College, Launceston, Tasmania*, 2023.[Under Examination]

13. BSI, "Guide to Measurement and Evaluation of Human Exposure to Whole-Body Mechanical Vibration and Repeated Shock." Sep. 30, 1987.

14. ISO, "Evaluation of human exposure to whole-body vibration - Part 3 : Evaluation of exposure to whole-body z-axis vertical vibration in the frequency range 0,l to 0,63 Hz." Jan. 05, 1985.

15. YARNALL, J N., HORN G W., and HUGHES M L., "Submarine Surface Abandonment Trials," *Naval Submarine Medical Research Laboratory, Groton, CT, USA, 50811/TR-2009-1273*, Nov. 2009.

16. JOUBERT, P N., "Some Aspects of Submarine Design Part 1. Hydrodynamics," *DSTO Defence Science and Technology Organisation, Fishermans Bend, Victoria, AUS, Technical Report DSTO-TR-1622*, Oct. 2004.

17. ITTC, "ITTC - Recommended Procedures and Guidelines: Resistance Test." ITTC, 2017.

18. JOUBERT, P N., "Some Aspects of Submarine Design Part 2. Shape of a Submarine 2026," *DSTO Defence Science and Technology Organisation, Fishermans Bend, Victoria, AUS, Technical Report DSTO-TR-1920*, Dec. 2006.

19. OVERPELT, B., NIENHUIS, B., and ANDERSON, B., "Free Running Manoeuvring Model Tests on a Modern Generic SSK Class Submarine (BB2)," presented at the *Pacific Int. Maritime Conference, Sydney, Australia*, Oct. 2015, p. 16.

20. COURDIER, M., LEONG, Z., DUFFY, J., BINNS, J., and CONWAY, A., "Numerical Validation of the Hydrodynamic Drag of a Surfaced and Fully Appended Underwater Vehicle," *Int. J. Marit. Eng., vol. 164*, no. 1, Jun. 2022

21. ITTC, "ITTC - Recommended Procedures and Guidelines: Model Manufacture, Ship Models." *ITTC*, 2017.

22. KUMAR, C, ET AL., "Boundary layer tripping study on a Generic Submarine Model in water," in *Proceedings of the 21st Australasian Fluid Mechanics Conference, 2018*, pp. 1–4. Accessed: Oct. 06, 2022.

23. MCLEAN, T., "An Investigation into the Resistance of a Surfaced Submarine.," *BEng Thesis, University of Tasmania, Launceston*, 2019.

24. BINNS, J R., and BRANDNER A P., "Yacht testing at the Autralian Maritime College," presented at *the International Conference - The Modern Yacht, London, UK, London, UK: RINA, 2003*, pp. 145–156.

25. ITTC, "ITTC - Recommended Procedures and Guidelines: Seakeeping Experiments Rev7." *ITTC*, 2021.

26. FRIEND, B., 'Exercise Black Carillion' 2013. 2013. Accessed: Jan. 20, 2023. [Photography]. Available: https://images.defence.gov.au/assets/Home/Search?Query=20131117ran8098578_057.jpg&Type=Filename

9. AUTHORS' BIOGRAPHY

Mathieu Courdier, AMRINA holds the position of Project Manager in the Defence Department at MARIN. His previous experiences include his PhD in underwater vehicle hydrodynamics at the Australian Maritime College, as well as working as a naval architect in naval defence and oil and gas companies.

Dr Zhi Leong serves as a Senior Research Fellow at the AMC. His research includes varied applications of Computational Fluid Dynamics and Experimental methods to investigate hydrodynamic challenges in Maritime Engineering, with a special interest in underwater vehicles.

Dr Jonathan Duffy is an associate professor at the AMC. He has particular interest in ship manoeuvring and ship behaviour in waves and has served in the ITTC manoeuvring Committee. Over his career he has developed strong strategic links with industry, facilitating research income and impact.

Dr Jonathan Binns is a professor of ship hydrodynamics and currently a group leader at DSTG for Naval Architecture and Platform Systems Analysis. His specialisation is in the hydrodynamic analysis of platforms performance. Previously he held various senior roles at the AMC and University of Tasmania where he led a range of maritime engineering projects.

Dr Alexander Conway is a researcher in hydrodynamics at DSTG for the Hydroacoustics Group.

APPLICATION OF SYSTEMS ENGINEERING (SE) WITHIN THE NAVAL ARCHITECTURE DOMAIN, WITH SPECIFIC APPLICATION OF MODEL BASED SYSTEMS ENGINEERING (MBSE)

M E Flexer and **B Bosley,** BAE Systems – Submarines, UK

SUMMARY

Future submarines are likely to face increasingly complex design challenges than previous decades; the introduction of automation, with the potential need for integration with unmanned vehicles, or other complex payloads and the need to shift to higher levels of availability will drive new and previously unexplored requirements. While Systems Engineering (SE) methods are widely used within the defence sector, the application of Model Based Systems Engineering (MBSE) is considered relatively new and untested, particularly through the whole life cycle of a marine vehicle. Outside Naval Architecture, MBSE is rapidly becoming an industry standard for complex system engineering, and is professed to offer advantages in quality, knowledge retention and agility in ability to make design change. The use of SE and MBSE is therefore expected to increase within the maritime defence sector, particularly to include increasing use outside the current early stages of acquisition, and into the detailed stages of system/capability design, integration and verification.

Questions arise around where the role of the Naval Architect sits within this 'new world order'; with some significant questions being asked such as should Naval Architects be considered Systems Architects, System Owners, or even a type of System Engineer, in addition can traditional methods such as the design spiral work within SE and MBSE processes?

While work is still ongoing in this area, this paper aims to examine some of the issues and potential benefits from a Naval Architecture perspective, including the need for both the detailed knowledge and whole boat view that Naval Architects may bring to the SE process.

In this paper it will be shown that while SE, specifically MBSE, bring new tools and more formalised language to system modelling, synergy with these ways of working can be seen with many Naval Architecture processes. This is seen as significant to Naval Architecture and specifically the training and future skills needed for submarine design going forward into the 21st century.

NOMENCLATURE

DODAF: Department of Defence Architecture Framework
MBSE: Model Based Systems Engineering
MBD: Model Based Design
MDA: Model Driven Architecture
MDAO: Multidisciplinary Design Analysis and Optimization
MODAF: Ministry of Defence Architecture Framework
OVM: Orthogonal Variability Model
PLE: Product Line Engineering
SE: Systems Engineering
SoS: System of Systems
SySML: Systems Modelling Language
SVN: Apache Subversion
UAF: Unified Architecture Framework
UML: Unified Modelling Language

1. INTRODUCTION

While Systems Engineering (SE) methods are widely used elsewhere within the defence sector, the application of Model Based Systems Engineering (MBSE) is considered relatively new and untested through the whole design life cycle at the whole boat and system level in Naval Architecture. Questions arise around what benefits there might be to Naval Architects, where they might sit within any new system engineering processes adopted and how their ways of working might be affected. This paper looks to address some of these considerations.

2. AIM

The aim of this paper is to relay some of the findings and conclusions by the authors following work carried out to investigate the impact of current and future use of MBSE in the context of Naval Architecture.

The work discussed in this paper is work in progress, and reflects upon the experiences and knowledge gained on MBSE from academic study and research. It represents the authors' opinions, and not those of BAE Systems.

3. TERMS OF REFERENCE

When looking at the subject of SE, specifically MBSE it is prudent to first define a number of the key terms used to avoid some of the misconceptions that might arise both internally within SE [1] and within Naval Architecture when discussing SE.

System
ISO 15288 [2] defines a system as a 'combination of interacting elements organized to achieve one or more stated purposes', while this fits comfortably with most engineered systems, in reality this definition of a system can be expanded to include any combination of interacting elements, whether they achieve a common purpose or not.

Architecture
Again using ISO 15288 [2] as a definition, architecture can be defined as 'fundamental concepts or properties of a system in its environment embodied in its elements, relationships, and in the principles of its design and evolution'. In terms of Naval Architecture this could be seen to translate into the key conceptual arrangements for the vessel being designed to meet Key User Requirements, for example the hull-form type (surface, sub-surface vessel, mono or multi hull), propulsion, control surface configuration, spatial arrangement, weapons, sensors etc.

Model
While there are a number of definitions of a model, in the context of this paper a model can be defined as a representation of a system, created with the purpose of increasing understanding, communicating, clarifying or designing aspects of the system or system of systems under consideration [3] [4] [5].

Systems Engineering (SE)
In the context of this paper, SE can be best described as an approach to engineering which is both transdisciplinary (i.e. crossing over disciplines) and integrative at all levels of a design, with the aim of assisting the realisation, use, and retirement of engineered product and systems, using scientific or management methods to aid this purpose [6].

Model Based Systems Engineering (MBSE)
Having defined SE as an integrative and transdisciplinary approach to the development of systems, MBSE can therefore be defined as a model based application of the SE approach, with the aim of assisting development of complex systems. However, unlike other types of modelling Naval Architects may be familiar with, this normally employs model based syntax such as UML or SySML, modelling frameworks (UAF, MODAF, DODAF etc.) and specialist toolsets.

Therefore, while MBSE can simply be defined as SE using models, it also commonly employs specialist modelling languages, toolsets and in some cases modelling frameworks.

4. MBSE METHODS IN THE CONTEXT OF NAVAL ARCHITECTURE

When considering MBSE in the context of Naval Architecture, some key questions need to be asked. i.e.

- What is MBSE?
- How does the MBSE or SE approach differ to those traditionally used by Naval Architecture?
- In the context of System Engineering, would a Naval Architect be classed as a System Owner or Architect or even type of Systems Engineer?
- What are the potential benefits of a MBSE approach to Naval Architecture?

- What do Naval Architects and Education Institutions need to do to prepare for wider adoption of MBSE within complex marine platforms?

4.1 CORE CONCEPTS

Having defined a number of key questions, we will explore some of these concepts.

4.1 (a) What is MBSE?

While the question around what is MBSE is defined in the terms of reference earlier in this paper, what it is and what it is not is worth exploring further. Returning to the terms of reference, it will be noted that many Naval Architects have been using some kind of model in their work throughout their career. In recent years this use of models may have increased even further with the widespread use of computer simulation (CFD, FEA etc.) and Computer Aided Design (CAD), it may also now include increasing used of additional design analysis or optimisation.

However, these type of modelling activities or methods are generally better described as Model Based Design (MBD), Model Driven Architecture (MDA) or even Multidisciplinary Design Analysis and Optimization (MDAO), not MBSE as sometimes misunderstood.

Therefore whilst working within Naval Architecture has shown there to be a number of touch points with these model based methods, MBSE can be seen to be quite different to other modelling Naval Architects are traditionally involved with.

Having previously defined MBSE as the model based application of a System Engineering approach, it is therefore worth expanding on what that means in practice. At its core MBSE gives the ability to create and maintain a conceptual description of a system through a multi layered interconnected model that describes: what the purpose of the system is, by whom it is used by, in what environment it needs to operate, what sub components it contains, what processes it needs to conduct in order to meet its goal, and how well it must do this.

Generally it can be seen to use specific languages (SySML/UML) but also specific design elements, typically split into three areas:

- **Requirements** – containing the goals, user needs and standards the system must fulfil

- **Behavioural** (Use case, activity, sequence diagrams) – describing at a functional level how a system is intended to work, its interactions with other system elements or its different conditions/states.

- **Structural** (Block, internal blocks) – describing the physical architectural arrangement of the system, sub-systems or systems elements, either at a black or white box level.

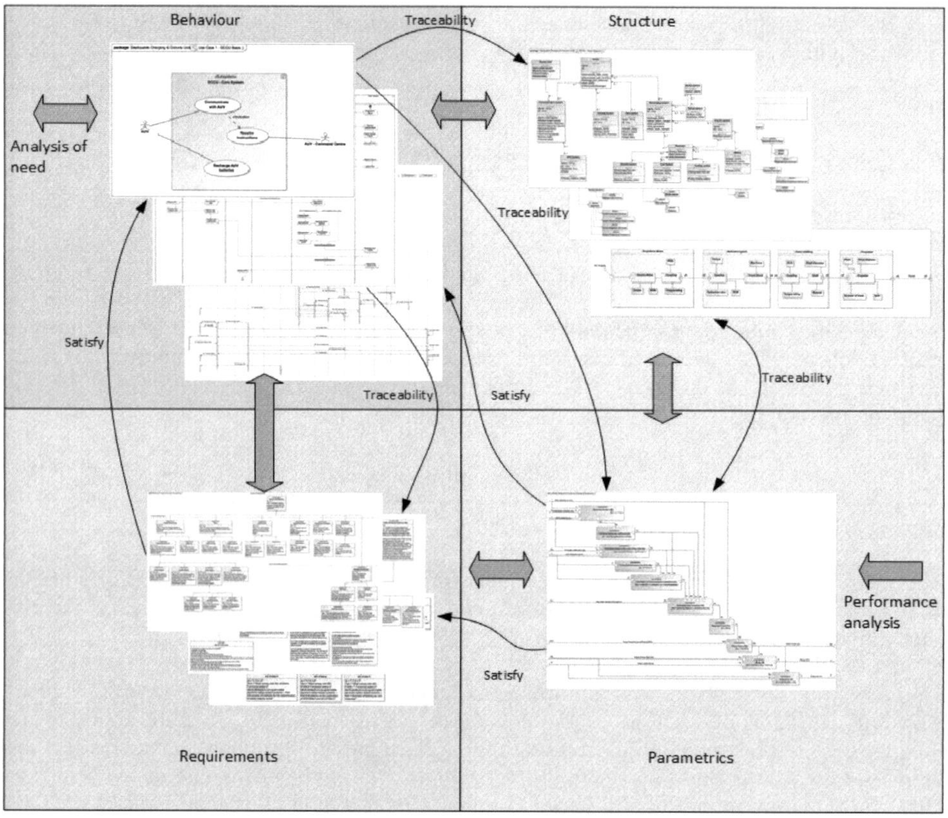

Figure 1: MBSE diagram types

Figure 1 shows some of the typical model types used in MBSE, with their interactions. Whilst the parametric view of the model is normally included in the structure group, they are generally linked to modelling the measure of performance and have been shown here separately to highlight their use in the model and system analysis.

4.1 (b) How does the MBSE or SE approach differ to those traditionally used by Naval Architecture?

When considering the application of SE and more specifically MBSE in the area of Naval Architecture, one of the first questions that may be asked is how this aligns with a Naval Architecture view of the development of a marine vehicle.

One of the key views still taught to Naval Architects is the design spiral, and while this has served Naval Architecture since 1959 [7], it has come under some criticism over recent years [8] [9], particularly in respect to its implied need to complete a full circuit of the spiral, fully completing each activity before moving to the next level of detail, it remains a key view taught to Naval Architects.

Therefore as SE and MBSE approaches become potentially more prevalent, how well does the work Naval Architects do align with a SE point of view?

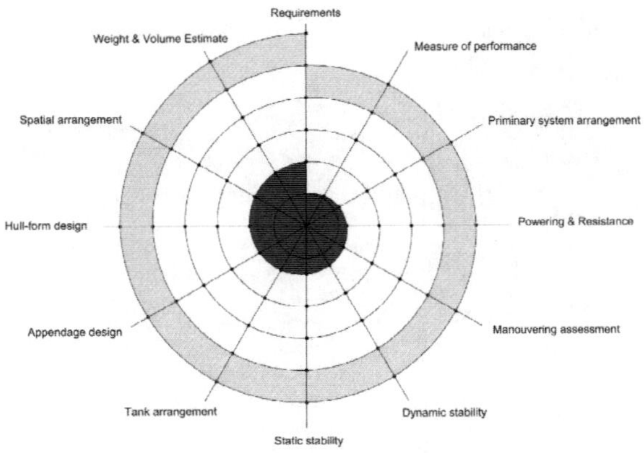

Figure 2: The traditional design spiral

One of the key views of a design lifecycle presented in a number of SE references [6] is the V diagram. In simplistic terms this shows the decomposition and integration of a design from left to right, starting with concept in the upper left of the V, before moving down the V to system architecture, then system design, before then starting to rise back up the V to verification, validation, integration and final in service operation.

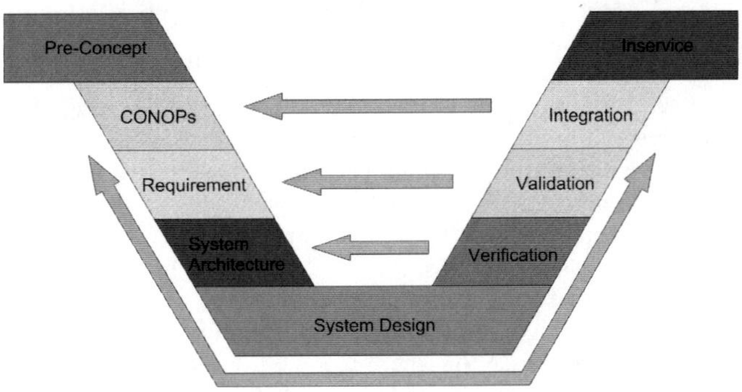

Figure 3: Typical V diagram

One of the key differences implied by the V diagram is that the design lifecycle moves in an almost linear fashion with one action being completed before moving on to the next (or waterfall in project management terms), not going back for further refinement or further details to be added.

However this implied progression is deceptive and in practice it is seen that the design will progress up and down the V, and in three dimensions (i.e. Z axis through the paper when looking at Figure 3) [10] as the design matures and levels of detail increase. Such that work may pass through the system architecture phase on to system design and on to verification, before returning to system architecture where any anomalies found during verification would be dealt with in change to the design.

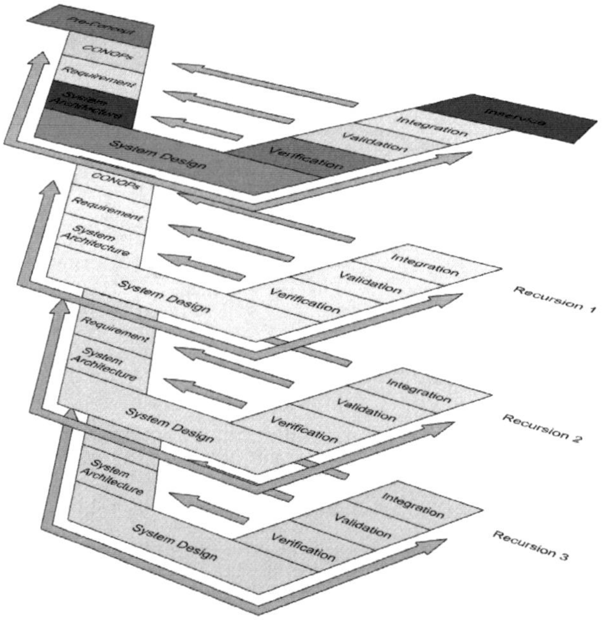

Figure 4: V diagram showing recursions

An alternative view shown in Figure 4 shows how the actual observed SE activity can be visualised as being broken down into a number of smaller Vs or 'recursion', as system maturity and levels of decomposition increase. Each level of recursion therefore feeding back information from the right-hand side (verification, validation, integration) back into the left-hand system solution (system architecture, system design, requirements) for further refinement of the system before proceeding back through the design cycle.

Given that it has been shown that the SE V diagram process is not actually the linear waterfall progression that may be perceived, then taking the step of stacking and expanding the recursions on top of each other into flattened 2D views, the argument is then that the SE process in practice starts to potentially become very similar to the spiral view we are more familiar with, but with different activity groupings.

4.1 (c) In the context of System Engineering, would a Naval Architect be classed as a System Owner or Architect or even a type of Systems Engineer?

Having looked at the high level process for MBSE and SE, and how similar or dissimilar this might be to methods used previously in Naval Architecture, one of the first questions when looking at the interaction between the two disciplines is to define where the Naval Architect might sit within the process.

Are they a System Owner (with discrete domain knowledge), an Architect (defining the overall design of a system, and all the main components etc.) or Systems Engineer (looking at the system at not only the high level System of System (SoS), but also the lower level specifics of how all components will work together)?

To understand what a Naval Architect is in the context of SE or MBSE, it may be prudent to attempt to first define what the role of a Naval Architect is in terms of ship building, particularly in the context of the defence industry.

While asking the question of what a Naval Architect does may result in a very different answer depending on the Naval Architect's current role, experience and training, some commonalities exists, including responsibility for areas such as hull-form design, hydrostatics, resistance/powering, concept development etc. However it is evident that these traditional roles are increasingly being accompanied by the need for Naval Architects to expand their role to bridge some of the gaps in the design process, helping define both physical and functional system integration, functional and non-functional requirements and detailed verification analysis, working at both a whole boat and system level to ensure the vessel delivers to its desired purpose.

Therefore while it could be argued that SE and specifically MBSE offers new tools and approaches to looking at the problems of designing, integrating and building marine vehicles via the use of system models, Naval Architecture traditionally fulfilled all three of these roles. Not only do Naval Architects commonly hold specific domain knowledge (control surface design for example), but also a high level view point of both the systems being employed to achieve the high level goal (steer the vessel) and the interactions necessary (hydraulics, structures, hydrodynamics etc.).

Table 1 shows some of the key similarities and differences identified between the three System Engineering roles with the Naval Architects role.

So having defined what SE and MBSE is, and where Naval Architecture might sit within this new discipline, what are the potential benefits of a MBSE approach?

Table 1. Comparison of the role of Naval Architects and System Owners, System Architects & Systems Engineers

	Similarities with the Naval Architect role	**Differences with the Naval Architect role**
System Owner	Detailed knowledge of functional requirements	Generally no responsibility for developing system maturity.
System Architect	Derive lower level requirements Own functions & assist or lead in architecture decisions. Creating multiple architectures that may meet the requirements, and helping down select an appropriate one.	Also responsible for specialist domain modelling (hydrostatics, hydrodynamic etc.)
Systems Engineer	Responsible for integrated functional performance. Responsible for a holistic whole boat design that meets customer requirements. Helping define stakeholder or customer needs. Helping derive lower level system requirements.	Generally not responsible for lifecycle management processes

4.2 POTENTIAL BENEFITS OF MBSE & SE FOR NAVAL ARCHITECTS

SE methods and MBSE have existed for decades, and therefore there are numerous examples of their application both inside and outside the maritime domain. Whilst there has been a number of examples on the use of MBSE for SoS design [11] and the transition from using architecture frames works such as UAF to more detailed integration of Systems or Sub-Systems [12], there remains limited published information on the actual application at a detailed system level within the marine domain, i.e. platform, level. Tepper [13], Kerns [14] and Wells [15] have described surfaced ship applications in the USA. Whilst Pearce [16] & [17] discusses the application of MBSE to submarine concept design. Common to all of these examples is the use of MBSE methods at the acquisition or concept stage of the engineering lifecycle. However what is less well published is the application at the later stages, particularly at the detailed system design phases.

However through examination of methods such as MBSE Grid [18] it has been shown that MBSE offers the potential for moving its use from the acquisition or early concept stages and have potential benefits to the Naval Architect in the following areas:

- Definition of functional behaviour
- Management of corporate knowledge/clear agreement of implicit requirements
- Traceability and links from functional behaviour to requirements
- Variant management

- Meta data and model integration

4.2 (a) Definition of functional behaviour

The use of MBSE enables the articulation of not just *what* a product needs to do, but *how* it does it and *how well* it needs to do it. The *how* is demonstrated through linked diagrams that join together functional behaviour, use cases and requirements. A particular type of model, called an Activity Diagram, is used to express the flow of functional behaviours of the system.

Using diagrams and models to capture this information is useful for a number of reasons:
1. It provides a **clear expression of what the vessel needs to do and why** – linking the functional behaviours to higher level use cases enables traceability to the fundamental requirement for that capability. Understanding functions at an early stage should result in less rework – and therefore less likelihood of causing imbalance in the design.
2. **It is solution agnostic**. Understanding what the goal is, rather than what has been done before will enable creative and innovative solutions to problems.
3. **It provides a baseline** - an agreement amongst parties early on about what it needs to do. When the project expands, a model based approach allows knowledge to be shared with a wider group, and it is not dependant on key individuals to oversee design areas. Having such a baseline ensures alignment at the beginning of the project on the scenarios designed for and assessed against. Otherwise, there is danger that the two do not align, and as such design change or new requirements appear later on in the design and cause rework. Key decisions and assumptions move on quickly in early stage design, and it is beneficial if this sits within a live model as the single source of truth.

4.2 (b) Management of corporate knowledge / clear agreement of implicit requirements

In the field of SE, implicit requirements are those that are not expressed as formal requirements by a customer. They are usually the things that are implicit in a customer's expectation of a working submarine design; the requirements that do not need to be stated, because by definition we expect a ship to do it. To derive these requirements, one has to have a detailed knowledge of how a ship operates, the impact of the environment and the physical effect on the platform and systems.

Within design programmes, Naval Architects, amongst others, are often called upon to have a broad understanding of the uses, context and fundamental functions required to fulfil the minimum expectation of a ship. They use this knowledge to support system design processes, including requirements generation and balancing the design when trade-offs are required. Derivation of the implicit requirements, therefore often falls to this group.

Creation of an MBSE model that houses these assumptions and implicit requirements becomes powerful when it comes to future designs. It enables a business to house its corporate knowledge within the model, and not within individuals. Particularly in the submarines domain, product knowledge is not readily available amongst the wider engineering population; businesses have to invest in individuals to "convert" them to this niche area of Naval Architecture. MBSE potentially offers a robust knowledge capture and transfer process.

It is possible to house the above in documentation, but in complex system design there are a number of drawbacks to this approach:

1. The complexity of the product lends itself towards a model – there will be multiple parallel processes and sub-processes that would be difficult to express adequately in words.
2. The design itself moves on and changes, particularly during concept design, and documents become outdated as soon as they are authored. The overhead of maintaining such documentation is challenging.

4.2 (c) Traceability and links from functional behaviour to requirements

When functional behaviour is defined, it is then possible to assign requirements to systems and integrated systems with Measures of Performance that have traceability to the function and back to the Use Case or higher level requirement. Thorough understanding of the origins of a system requirement are invaluable to trade off decisions and concession arguments. This is particularly relevant to Naval Architects due to our role in both physical and functional integration of systems, and the need to satisfy protagonist requirements.

Naval Architects are often asked to assess the impact of reducing functional performance, and one of the main difficulties in assessing this is identifying and down selecting the bounding or most important case or cases. A readily accessible mapping to where the each aspect of the requirement comes from can be important.

If there is a design constraint such as weight or space that prohibits the system meeting its requirements, direct mapping to Use Cases and Behaviour Diagrams is advantageous in quickly gaining understanding of the impact of the trade of functional performance against practical constraints of the design. The same may be seen when trades need to occur between systems – in the example of weight, where a functional performance reduction is not acceptable, it may be possible to look to another system to reduce weight, and then mapping both these systems to the higher level Use Case enables understanding of what limitation you are driving into the platform's capability.

4.2 (d) Variant management

One of the other key challenges facing ship design and build in the future is to how to deal with variation between designs, both within the same class (change), new classes (re-use), or in service (future upgrades etc.). Current methods for controlling these changes is labour intensive, inefficient and prone to error or omission i.e. how do we know a change has been flowed across all boats in a class or change to one vessel recorded correctly?

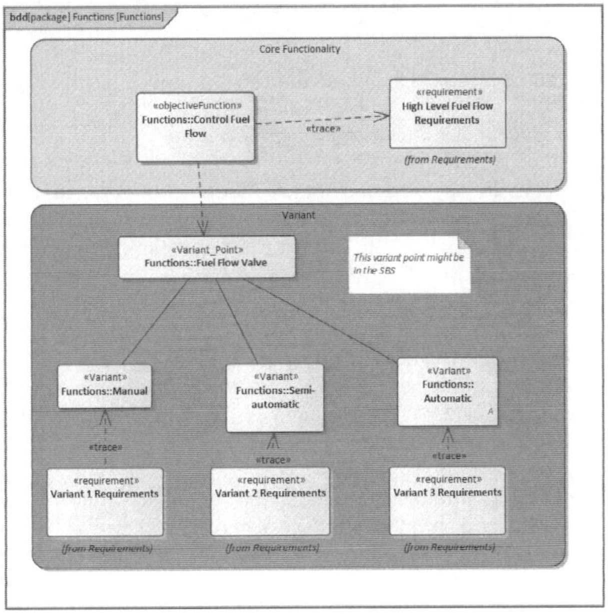

Figure 5: Typical 150 % model showing variant point concept.

This is an area where the move over to a more model based approach to SE may aid control and management of variability, either through a managed approach to cloning or full move to a Product Line Engineering (PLE) approach [19], including the development of models with all differences contained within them (150% models), combined with separate models defining the variations and variant points within the design (OVM). Alternative methods include managed cloning using version management methods such as SVN or GIT, Variant Modelling with SysML (VAMOS) [20], modularisation and other methods developed in software development. Some of the further benefit of PLE and variant modelling is explored in 'Model-Based Product Line Engineering to Plan and Track Submarine Configuration' [21].

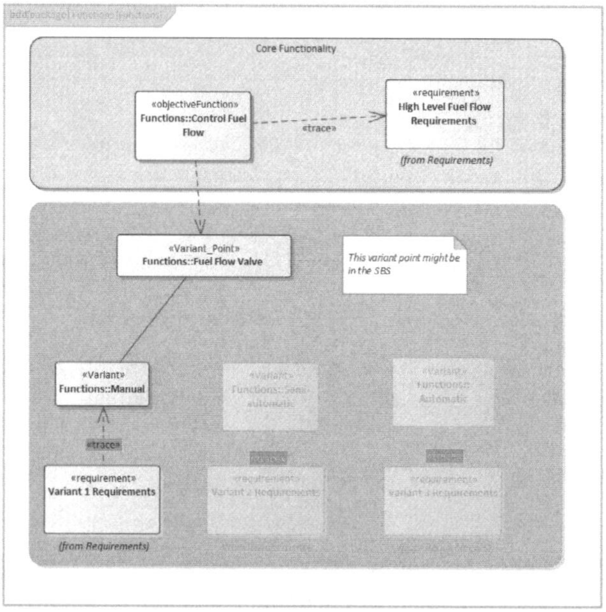

Figure 6: Typical 150% model filtered to show single variant

4.2 (e) Information management, meta data and model integration

Another significant opportunity exists in the use of the MBSE model as a centralised repository for properties (such as mass, volume, IP rating) and performance data. Performance models at the whole boat may contain hundreds of inputs, with the source of the data originating from many different system performance models, CAD models, documents and databases.

Historically each model owner will engage separately with the owner of the relevant document to gain access and coordinate the production of the required information (see Figure 7 below). The effort is significant, open to error and requires significant obedience and forethought by the engineers to engage with one another regularly and update the model owners with any changes. Particularly at the early stages of design, when design change is constant, this becomes challenging.

If the central MBSE model were to hold the data, a single centralised repository would exist for the various model owners to extract from, and commit data to (see Figure 8).

The power of MBSE is such that the model can house metadata that links not just the source of the data but also where it is being used, creating an instant map of who is dependent on data if and when it changes.

An initial investigation of this capability has been conducted using Enterprise Architect (EA) models to store data, and then extracting this data to Resistance & Propulsion (MATLAB), Stability & Manoeuvring (Paramarine) models through intermediate interfaces. Then reversing this process via extracting data from these models to return to EA. This enables all three performance models to use common data, and assurance is gained that a single source of truth is maintained.

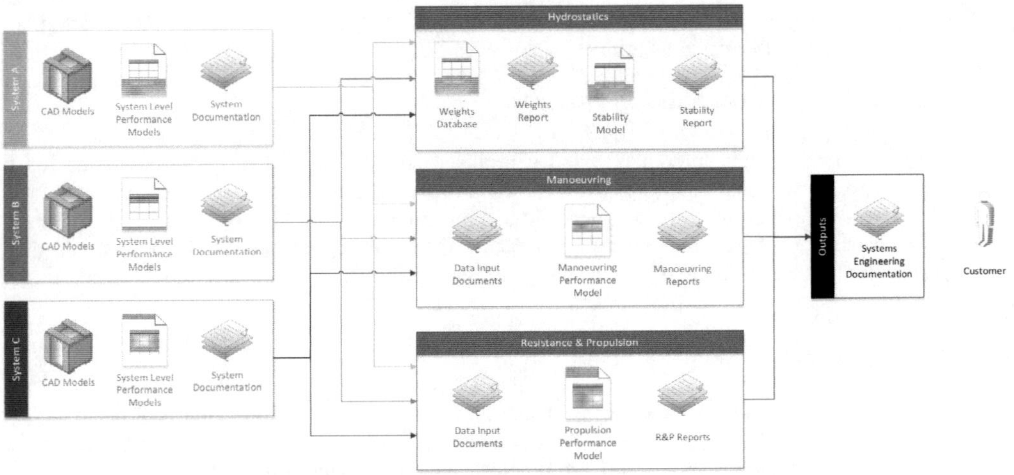

Figure 7: Current stove-piped document based system of work

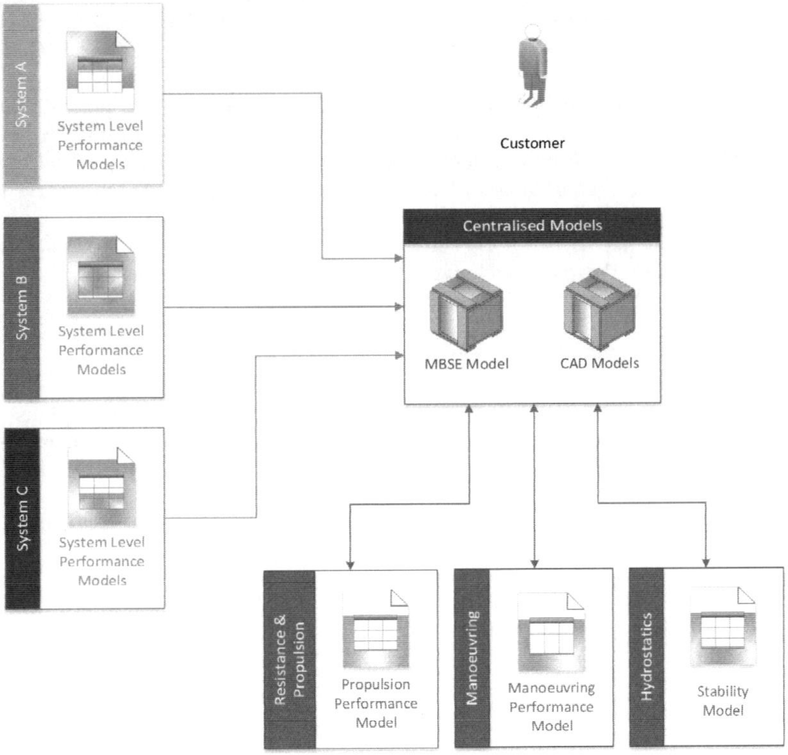

Figure 8: Model centric approach to data management

4.3 FUTURE DEVELOPMENT

4.3 (a) Training and education

Whilst MBSE has been deployed on many successful programmes in a variety of domains, there remains a limited uptake and knowledge base in the Naval Architecture community. It has been witnessed that there can be misunderstanding in what MBSE is and what it is not. Engineers have a tendency to associate it with the type of performance and analytical modelling they have experienced in their careers. Interacting with MBSE models also requires an understanding of the languages behind them, and the terminology of SE. Sometimes the terminology used in SE can be similar but very different in definitions and meaning to that used by Naval Architecture, for example the words Model, Architecture and Structure used elsewhere in this paper, hence producing confusion. Improvements can be achieved by mass training and education in the workplace, but could also be assisted by Further Education.

4.3 (b) Model integration and the use of meta data

Initial demonstrations of data interchange and centralised storage have identified areas that need further consideration as to the automation of the exchange, and the protocols required to establish quality and control. To enable wide use of the capability, there needs to be fundamental trust in the software and processes that the source of the data is configured, managed and up to date. The process also needs to be bought into by a wide group of stakeholders, otherwise partial uptake will not see the benefits come to fruition, and risks duplication of effort.

5. CONCLUSIONS

While MBSE can be considered a maturing discipline within large scale defence projects, within Naval Architecture there are already many touch points, however it is not to be confused with other modelling techniques traditionally used within Naval Architecture and engineering. It can also be concluded that there are many similarities existing between the roles of Naval Architects and Systems Engineers.

Synergies have been shown between the Design Spiral and the SE V Diagram, with it being demonstrated within this paper that the V diagram could be seen as an alternative view of the Design Spiral.

Returning to some of the problems presented at the start of this paper, such as increasing complexity of the platforms and payloads, management of change, need for higher availability of platforms, integration etc, it could be concluded MBSE could offer new tools or methodologies complimenting Naval Architecture current ways of working, however further training of Naval Architects would be needed to achieve this.

Of the many different benefits MBSE offers Naval Architects, these remain to be proven in large scale ship design projects throughout the design lifecycle. Work therefore needs to continue to investigate and demonstration the benefits SE and MBSE at the detailed and whole boat level.

6. ACKNOWLEDGEMENTS

The authors would like to thank Michael Wilkinson (Chief System Engineer, BAE Systems), David Dempster (Systems Engineering, BAE Systems), Toby Peacock (Senior Technologist, BAE Systems) for their input and guidance in development of this paper.

7. REFERENCES

1. SMITH, G M., 'Indentifying the Key Misconceptions in Systems Engineering', 2021.

2. ISO, BS ISO/IEC/IEEE 15288:2015 - Systems and software engineering - System life cycle processes, 2015.

3. BELLINGER, G., 'Modeling & simulation: An introduction', *Mental Model Musings*, 2004.

4. DORI, D., ‘Object-Process Methodology: A Holistic System Paradigm’, 2002.

5. OMG (Object Management Group), 'MDA Foundation Model', 2010.

6. INCOSE, 'Systems Engineering Handbook', *Fourth Edition* 2015.

7. EVANS, J H., 'Basic design concepts', *Journal of the American Society for Naval Engineers*, 1959.

8. RADOSAVLJEVIC, D., 'Simulation-driven ship design', *Siemens*, 2022.

9. PAWLING, R., ANDREWS, D. and PERCIVAL, V., 'A Study into the Validity of the Ship Design Spiral in Early Stage Ship Design', *University College London*, 2019.

10. FORSBERG, K. and MOOZ, H., 'The Relationship of System Engineering to the Project Cycle', 1994.

11. ROGERS, S W., 'MBSE delivers significant return on investment in evolutionary development of complex SoS', 2021.

12. MORKEVICIUSIT, A., 'From UAF to SysML: Transitioning from System of Systems to Systems Architecture', 2021.

13. TEPPER, N., 'Exploring the use of Model-Based Systems Engineering (MBSE) to develop Systems Architectures in Naval Ship Design', *Massachusetts Institute of Technology*, 2010.

14. KERNS, C., 'Naval Ship Design and Synthesis Model Architecture Using a Model-Based Systems Engineering Approach', *Virginia Tech*, 2011.

15. WELLS, B H., 'Applying Systems Engineering to Naval Shipbuilding', *INCOSE*, 2006.

16. PEARCE, P., 'Introducing MBSE to a Submarine Concept Design Team', *Deep Blue Tech*, 2013.

17. PEARCE, P., and HAUSE, M., 'ISO-15288, OOSEM and Model-Based Submarine Design', *SETE APCOSE*, 2012.

18. MORKEVICIUSIT., Et Al, ' MBSE Grid: A Simplified SysML-Based Approach for Modeling Complex Systems', 2017.

19. LE PUT, A., 'System Product Line Engineering Handbook', 2015.

20. WEILKIENS, T., 'Variant Modelling with SySML', 2012.

21. HAUSE, M., and HALLETT, J., Model-Based Product Line Engineering to Plan and Track Submarine Configuration', 2016.

8. AUTHORS' BIOGRAPHY

Morgan E Flexer, CEng, MRINA, currently holds the position of Engineering Manager at BAE Systems. He is responsible for the Weights, Infrastructure, Seamanship and Hydrostatics functions within the Naval Architecture Department. His previous experience includes running the Design & Development function within the Naval Architecture Department, leading on hull-form development, concept development and Research &Technology etc. Morgan currently in his final year of a 3 year part-time MSc in Systems Engineering at Loughborough university. Prior joining BAE Submarines, Morgan comes from working within the commercial & military small craft sector of Naval Architecture, having worked at Camarc Design and consultancy at Amgram Ltd.

Becky Bosley, CEng, MRINA, currently holds the position of Engineering Manager at BAE Systems. She is responsible for the management of the Hydrodynamics team within the Naval Architecture Department, a team of specialist engineers in fluid dynamics, manoeuvring analysis and submarine hydrodynamic design. Her experience includes submarine manoeuvring analysis, control system functional design and submarine sea trials.

DELIVERING SUBMARINE AVAILABILITY THROUGH DESIGN

R D Millard and **A E Walchester**, BMT, UK

SUMMARY

Submarine availability is critical to being able to deliver capability when required, and submarine operating navies seek to maximise operational availability, i.e. for mission tasking at sea and minimise non-operational time, i.e., for planned or unplanned maintenance, inspection and training.

The importance of availability is brought home in the capability planning stage, where the planned availability for a given submarine design influences required fleet numbers and procurement costs. Submarine availability is influenced by a variety of factors including (1) the submarine design itself and (2) the wider submarine enterprise engaged in support i.e., infrastructure, personnel capacity and capability, supply chain aspects.

It is necessary for designers to understand these considerations in the early stages of the submarine design process and to identify design features that maximise the chance of meeting the defined availability targets. This can be extremely difficult to address with the risk that design decisions made early can quickly lead to "locked in" costs only realised in the future in-service support phase.

This paper revisits BMT's research into submarine availability and takes a fresh look at how availability requirements influence the design. This paper proposes that by driving maintainability features into all aspects of the design process, and by developing the support model from an early stage in parallel, it is possible to maximise confidence across the enterprise that the defined availability target will be met.

1. INTRODUCTION

To ensure that scheduled maintenance periods remain on track, and that refit delays do not trigger a cascading effect which impacts the availability of the flotilla, the maintenance and support requirements of a concept submarine need to be considered at the earliest stages of the design process. The greatest requirements for submarine supportability come long after the first boat in a class has been delivered to the customer, and it is at this point that it is the hardest to make the level of changes required.

This paper revisits BMT's research into submarine availability [1][2] and takes a further look at how availability requirements influence the design.

1.1 DEFINING AVAILABILITY

An *Available* submarine is one that is materially capable of carrying out an assigned mission or patrol, whilst its *Availability* is the proportion of time for which a submarine, or a flotilla of submarines, is available.

It is important to note that the availability of the submarine in the context of this paper is only one part of the equation – a submarine's readiness for mission tasking at sea depends on factors beyond its material availability such as the availability of the crew and the fighting components of the submarine. In recognition of this, the phrase *Material Ready* is defined as meaning that the submarine is not conducting planned maintenance and is not encumbered by defects that prevent it from proceeding to sea and conducting its mission, whilst *Material Capable* means that the submarine is ready for operations with a 'worked-up' crew. Whilst there are clearly many factors involved, submarine availability i.e., being Material Ready, remains a fundamental element of ensuring a submarine is Material Capable when it is planned to be so within the fleet operating schedule.

1.2 CHALLENGES TO AVAILABILITY

Maximising submarine availability is a challenge faced by navies throughout the world. To combat increased costs associated with build and through life operation and a drive to reduce defence spending as a percentage of GDP [3] over the past decade, submarine flotillas have typically been reduced to the minimum acceptable hull numbers that can still achieve a government's defence priorities. This reliance on fewer submarines means that the current global trend for submarine availability is becoming increasingly challenging, rather than improving [4].

Slow development of specifications and design cycles, increases in build duration, delays to replacement classes, minimum hull numbers, and the desire to maintain a continued submarine capability, mean existing submarines are increasingly turning to extended commissions (time between major maintenance periods) or life extensions (which lead to additional commissions) to meet the availability targets [4] required to achieve the defence priorities. Subsequently, the major maintenance work packages required to sustain ageing flotillas safely during life extensions are increasing in complexity, duration, and cost, drawing heavily on the limited submarine infrastructure and Suitably Qualified and Experienced Personnel (SQEP) with the Submarine Enterprise. Any delays to these major undertakings exacerbate the already fragile availability of the flotilla and have knock on impacts to subsequent submarine programmes.

The desire for more effective and flexible submarines often means that each successive submarine class is fitted with increasingly complex technologies and systems to introduce additional capabilities whilst the number of submarines within the flotilla reduces. A consequence of this is introduction an additional obsolescence management burden on the enterprise. An example of this being the addition of highly complex combat systems such as the Tomahawk Land Attack Missiles (TLAM) added to the Swiftsure and Trafalgar Class boats the late 1990's [5], and recently contracted to be upgraded on the latest Astute class submarines [6].

With submarine numbers decreasing and major equipment becoming ever more bespoke, the number of suppliers are reducing; there is simply not enough reward in manufacturing unique boat sets of equipment for one class of submarine when compared to the standardised volumes that could be made for other industries. This diminishing supply chain further exacerbates the availability of spares, experience and affordable equipment.

Emergent defects, increasing refit times and a lack of ready to use spares has seen these minimum submarine numbers stress tested, resulting in low submarine availability. To overcome these challenges requires continuous improvement in many of the factors identified in the RAND [7] and Coles [8] reviews as well as those observed by BMT including fixed infrastructure constraints and maintaining a functioning Submarine Enterprise.

2. MAINTAINABILITY IN DESIGN

2.1 THE MAINTENANCE CYCLE

A submarine will suffer from degradation of its systems and overall performance within a maintenance cycle, reducing the reliability of the boat as a whole. The undertaking of the necessary maintenance impacts directly on availability. Degradation may be gradual, the result of a range of Age-Related Degradation Mechanisms (ARDMs) such as the breakdown of protective coatings and corrosion, or it may arise suddenly from operational incidents or events.

Preventative and corrective maintenance actions must be undertaken so that degradation effects do not compromise the planned availability of the submarine through-life, and to ensure that the original design intent is maintained or updated, and that the submarine is capable and safe to operate, as illustrated in Figure 1.

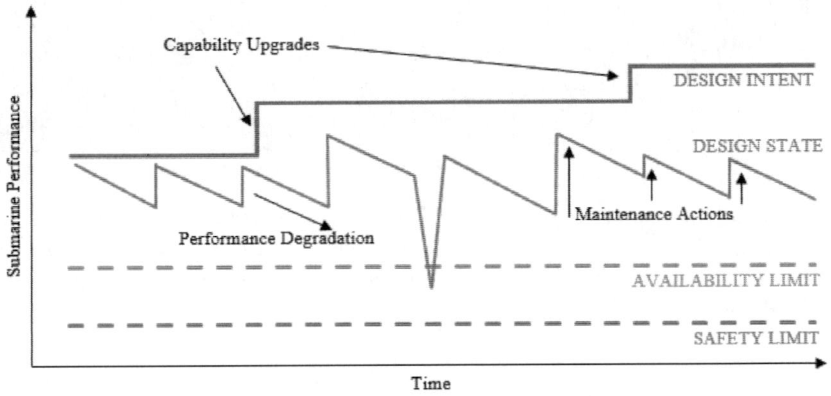

Figure 1: Variation in Submarine Performance with Time

Driving maintainability features into all aspects of the submarine design process from the earliest stages onwards will bring benefits throughout the life of the flotilla. This requires a progression in the approach from one where the design is targeted towards facilitating build and reducing the initial purchase cost, to one where through-life supportability and ultimately availability is given greater significance [7].

2.2 DEVELOPING THE SUPPORT MODEL

To measure the benefit of design changes intended to improve availability, BMT's research has identified that quantifying any benefit is dependent on knowing the tasks which lie on the critical path within the key activities within Maintenance Upkeep and Operating Cycle (MUOC). The MUOC is a time-based definition of the availability of a submarine and when combined when with the MUOC of the other submarines in the class will provide a view of the class availability. Problems arise when the planned availability of one submarine is not achieved, the whole submarine programme is impacted, introducing more unplanned unavailability.

A flotilla-wide example was developed for BMT's WYVERN® SSK concept design, as shown in Figure 3. BMT produced the WYVERN concept design in 2017 to investigate the feasibility and capabilities of a small SSK [9]. The submarine design (shown in Figure 2) has a submerged displacement of 860 tonnes, 4 torpedo tubes and an Air Independent Propulsion (AIP) system.

Figure 2: The BMT WYVERN® SSK Concept Design

WYVERN is an in-house design and not designed to the requirements of a specific nation, which enables us to investigate the impact of major design changes on the through-life availability of the theoretical flotilla.

The approach used in WYVERN's support model offers some significant benefits; delivering platform availability across a fleet of four that varies between two and three platforms at any given point in time. With at least four Extensive Maintenance Periods throughout the submarine's life, there is greater opportunity to rectify the defects and degradation accumulated over an operating period and smooths the distribution of maintenance workload for the support enterprise. However, these benefits can only be realised if the platform is designed in such a way as to enable extensive maintenance over a short duration – i.e., the design must fit the support model.

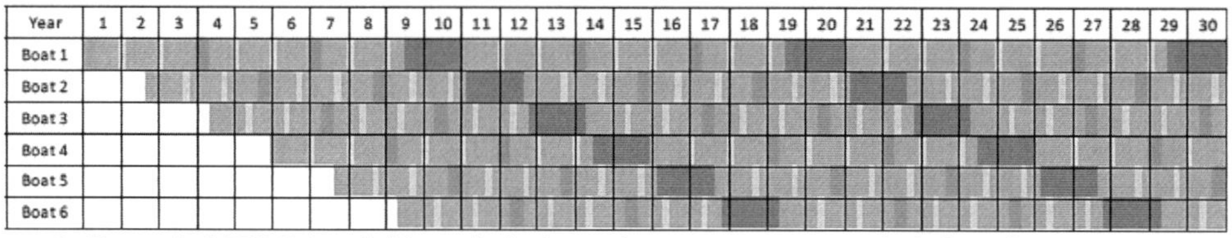

Figure 3: Example MUOC (available periods shown in green; support periods in yellow, orange and red)

For an existing class, this information should be readily available however in practice such a model is extremely challenging to develop at the concept design phase due to limitations of maintenance data and because of the extent of deviation in future designs from previous classes. The challenge increases when the impact of influences outside of design are considered.

2.3 ADDITIONAL INFLUENCES ON AVAILABILITY

The design of the submarine is only one of several important and significant influences on the availability subsequently achieved. The key 'non-design' influences considered are;

- **Concept of Operation (CONOPS)**: The Concept of Operation (CONOPS) defines the way the submarine is operated, and aspects such as continuity or irregularity of use have a significant impact on the submarine's continued ability to maintain its desired operational posture. It is from this Concept of Operation that all other requirements cascade down, ultimately driving the availability targets and supportability requirements.
-
- **Crew & Training**: The technical complexity of a modern submarine relies on sound design and efficacy of support; the Crew and Training are all a key part of achieving the required availability. The submarine design and support arrangements can be of the highest quality, but without a motivated Suitably Qualified and Experienced (SQEP) crew, shortfalls (such as unsafe operation and reduced equipment reliability) will be experienced.

- **Submarine Enterprise**: The Submarine Enterprise, taken to mean the combination of all of the individual contributors to the delivery of availability and enhancement of capability (i.e., the maintenance policy/philosophy and management process, logistics support, infrastructure, organisation, etc.), is highlighted as critical. This position is supported by good evidence of the significant benefits to the Australian Submarine Enterprise that have resulted from changes implemented through recommendations of the Coles Review [8]. The key stakeholders which form the Submarine Enterprise are captured in Figure 4

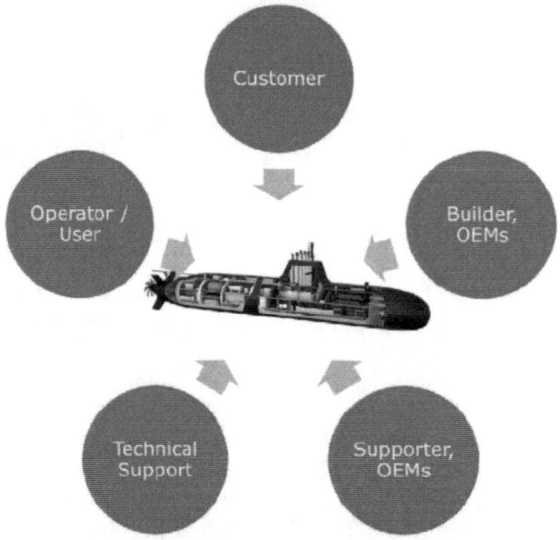

Figure 4: Key elements of the Submarine Enterprise

- **Supportability**: A submarine's support model is both greatly dependent on, and drives, a wide range of factors including.
    - Achieving the CONOPS
    - Infrastructure requirements (Can an operating nation provide the necessary maintenance facilities to support throughout the life of the flotilla?)
    - supply chain, human factors, industrial workforce resources and the
    - capability and skills of the people.

All these elements have to work together in harmony according to the plan represented by the support model, which also has to align with the key characteristics of flotilla size, design life, commission length, build 'drumbeat', and duration of the various types of maintenance period. Hence, the act of starting with an embryonic support model during the earliest stages provides the design team with a basis from which to develop a plan that aims to harmonise all these elements as the design evolves. The importance of the management a through-life support model (the MUOC) at the earliest design stages are highlighted by Schank et al in their study on the lessons learned from the US, UK and Australian submarine programmes [7].

2.4 AVAILABILITY LEVERS

To investigate opportunities for design options in greater detail from an availability perspective, BMT have applied three key availability 'levers' that represent the linkage between maintainability and availability:

- **Reduce Planned Maintenance**: Increase in the interval between planned maintenance activities by addressing constraints on the life of components. Improving the longevity of equipment and systems as well as aligning regular maintenance activities.
- **Reduce Days Lost to Defects**: Reduction in the probability of defects occurring on the submarine during fleet time by improving equipment and system reliability.
- **Reduce Maintenance Overrun**: Reduction in the duration of planned maintenance activities by configuring the submarine to allow more rapid maintenance/repair. Improving the process for planning and managing maintenance/repair activities to reduce the timescale, minimise conflicts, be prepared for emergent work and therefore improve schedule adherence.

The benefit of each technology or design change can be measured by its ability to improve maintainability, and in turn its contribution to the class in meeting the availability target. The relationship between these three levers and the wider enterprise considerations is shown in Figure 5.

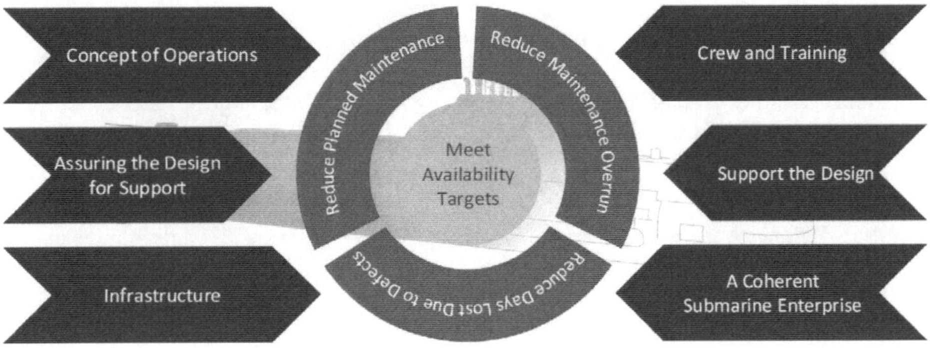

Figure 5: Availability Levers

2.5 AXIOMS

Five key axioms (statements that are regarded as being self-evidently true) have been identified [10] that 'can be used today to influence the conceptual design of equipment':

- **Simplicity**: The number of parts should be kept to a minimum and undue complexity avoided. Each component should have a clearly defined function and unnecessary or superfluous components removed; For example, rationalising and reducing the number of tanks in a design to reduce the inspection and maintenance burden.

- **Parts/components**: The weights of components, their shape, surface texture, space for tool access and so on are key factors to achieve good maintainability. The use of standardised components, especially fastener sizes, leads to high maintainability.
- **System environment**: The environment in which maintenance is carried out should be controlled to facilitate maintenance actions; For example, the design of the external structure to safely allow for in-water inspection and repair of equipment and structures outside the Pressure Hull.
- **Assembly/disassembly**: The sequence and combinations of actions to disassemble and reassemble components should be as simple as possible. This may be achieved through improving the modularity of systems to facilitate removal for maintenance ashore, or rapid swap-out.
- **Identification**: Parts and connections should be clearly identified, and adjustments/calibration facilitated.

The five axioms are the primary mechanism for improving the availability of a candidate piece of equipment or system and therefore to provide the justification for its inclusion in the submarine. The key question is how can we as submarine naval architects improve submarine availability by using these five axioms?

3. DESIGNING TO IMPROVE AVAILABILITY IN PRACTICE

3.1 PREVIOUS APPLICATION

Recognising the availability challenges faced by submarine operating navies, BMT has undertaken to understand the major influences on submarine availability and to assess the impact to availability that potential solutions would have on the design of existing or future submarines.

As presented previously [2], the WYVERN model has been used to investigate the impact of introducing a Collins Class style hull cuts at a concept design stage, which would enable the entire support model to be built around this design feature for maximum benefit in terms of maintenance.

The hull cut model (shown in Figure 6), named 'WYVERN Splice', is a significant and fundamental design feature that is aimed at enabling multiple circumferential hull cuts during every Extensive Maintenance Period. The objective being to enable by design from the outset, reassembly and commissioning within days rather than months.

Figure 6: WYVERN Splice

This hull cut method could provide increased levels of access for maintenance whilst minimising cost penalties and risk to the schedule that usually accompany such an approach. The design feature when coupled with the detailed consideration of Zonal Design processes this feature could significantly improve the confidence in meeting the 18-month duration of the longest maintenance period in WYVERN's support model.

3.2 PROCESS DEVELOPMENT

Learning from the experience gained during the development of the WYVERN Splice, BMT has built upon earlier proposed methods [1] for evaluating new design concepts and technologies early in the design process, to identify those changes that are most likely to improve the availability of future submarines. The overall assessment process is summarised in Figure 7. Taking a novel design idea or modification, the process is then used to assess them using the criteria embedded in the levers and axioms, and ultimately provide a ranking for cost and benefit.

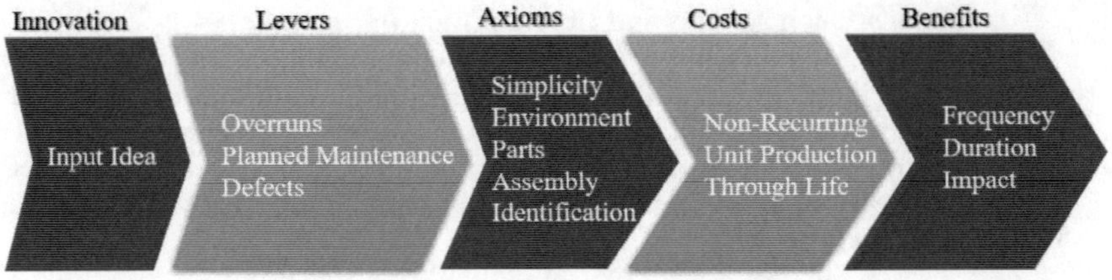

Figure 7: Design Concept Assessment Process

The ranking methodology applied to the candidate design concepts and technologies, accounts for the criticality of the maintenance activity that the design feature is trying to improve and therefore avoids investing in design features that are unlikely to sit on the MUOC critical path, and hence are unlikely to improve availability.

The criteria used to consider the cost/benefit aspects are:

- **Cost**
  - **Design Cost**: Will this technology feature substantially change the design effort of the design team?
  - **Unit Production Cost (UPC)**: Will this technology feature make a change to the cost of materials, fabrication, assembly, testing and commissioning?
  - **Technology Development Costs**: Will this technology feature a novel or new technology that requires to further development before it can be employed?
  - **Infrastructure Support Costs**: Will this technology feature require a more/less complex support solution and hence change in through life cost?

- **Benefit**
  - **Frequency**: Will this technology feature change how often the survey maintenance / repair / upgrade activity is conducted?
  - **Duration**: Will this technology feature change the duration of the survey /maintenance / repair / upgrade activity?
  - **Impact**: Will this technology feature change the complexity of the survey / maintenance / repair / upgrade activity?

Starting with a large selection of tasks which typically lie on the MUOC critical path; technologies and design concepts are then considered and refined through iterative assessments: Ideas for improvements can be grouped into the 'lever' group where they will have the greatest impact and are represented in a grid format, as shown in Figure 8.

Figure 8: Grouping of candidate ideas

3.3 EXAMPLE USAGE

Taking a single idea in greater detail through the initial assessment gives a demonstration of how the first pass selection is made. The high-level concept of improving access to WYVERN through the development of removable maintenance modules has been selected from the design concepts in Figure 8 as an example:

- Step 1 - **Consideration of Axioms and Levers for greater modularity**
  - Introduction of structural elements which would facilitate better access to remove equipment quickly and safely for repair ashore. This would benefit both the potential provide rapid capability changes, and for the replacement of defective equipment alongside.
  - The introduction of modular elements at the design stage serves to reduce the length of docking periods by increasing the number of maintenance activities which can be achieved without a hull cut.

- Step 2 - **Key cost/benefit considerations brought by increased modularity in submarine design**
  - Unit Purchase Cost could be expected to remain similar due to:
    - A cost increase due to the procurement and welding of additional steel.
    - A cost reduction due to the modular systems being expected to reduce build and commissioning times with an associated saving.
  - Through Life Cost could be expected to reduce overall due to:
    - Quicker equipment removal and replacement with less work in way.
    - Greater upkeep by exchange of problematic equipment leading to a potential reduction in training requirements for Ship's Staff.

4. CONCLUSION

The key challenges to submarine availability considered in this paper are; the development of submarine flotillas to operate with minimum hull numbers, competing interests driving a reduction in available defence budget; the strength of the supporting submarine enterprise; the maintainability of the design and the quality of the support model. Furthermore, the experiences of submarine operating navies this millennium, and in particular the UK, US, and Australia, has demonstrated that when the supportability and enterprise factors which underpin submarine availability are not actively worked on, the availability the flotilla and the wider submarine support network suffers.

The design principles presented in this paper highlight the importance of focussing design efforts for maintainability in the areas which will have the greatest impact on the MUOC cycle, which in turn will provide the best return on availability. If a submarine flotilla is to achieve its intended availability target, these considerations need to be kept at the fore throughout the design process. The designer can identify candidate design features which can contribute to improving submarine availability by using the design 'levers' of; reducing planned maintenance through more robust systems, reducing days lost to defects through improved system reliability and to reduce maintenance overrun by improving accessibility.

In the process described in this paper, the candidate ideas are then subjected to the five axioms of; simplicity, parts/components, system environment, assembly/disassembly, and identification before undergoing an initial cost/benefit analysis. This first-pass approach has enabled the availability impact of novel design concepts to be considered and can readily be applied to the preliminary stages of submarine design programmes.

Whilst as a concept this process appears straightforward, building the approach in to a submarine enterprise from start to finish is an exceptionally large challenge.

5. ACKNOWLEDGEMENTS

The authors would like to thank their colleagues at BMT who have supported the writing of this paper. In particular, thanks go to Ian Duncan, Ross Eddy, David Morris, Richard Hemsley, Angus Watt and Paul Buckton.

6. REFERENCES

1. SPINNEY, D., MORRIS, D., 'Design Methods for Challenging Submarine Availability Targets', *UDT, BMT*, 2016

2. SPINNEY, D., MORRIS, D., 'Designing for Support to Meet Challenging Submarine Availability Targets', *BMT*, 2017

3. Government expenditure on defence - Statistics Explained (europa.eu) https://ec.europa.eu/eurostat/statistics-

explained/index.php?title=Government_expenditure_on_defence#Evolution_of_.27defence.27_expenditure_over_1995-2021 (Accessed 04/2023)

4. SCHANK, J F., LACROIX, F W., MURPHY, R E., IP, C., ARENA, M V., KAMARCK, K N and LEE, G T., 'Learning from Experience – Lessons from the U.S. Navy's *Ohio*, *Seawolf* and *Virginia* Submarine Programs', Vol 2, RAND Corporation, 2011.

5. HENNESSY, P., JINKS, J., 'The Silent Deep – The Royal Navy Submarine Service since 1945', *Penguin Random House*, 2016

6. Submarine Service's Tomahawk missiles receive £265m revamp (mod.uk), https://www.royalnavy.mod.uk/news-and-latest-activity/news/2022/may/31/20220531-submarine-services-tomahawk-missiles-receive-265m-revamp, (Accessed 04/2023)

7. SCHANK, J F., LACROIX, F W., MURPHY, R E., IP, C., ARENA, M V., and LEE, G T., 'Learning from Experience – Lessons from the Submarine Programs of the United States, United Kingdom and Australia', *Vol 1, RAND Corporation*, 2011.

8. COLES, J., GREENFIELD, P., FISHER, A, and DAVIES, S., 'Study into the Business of Sustaining Australia's Strategic Collins Class Submarine Capability – Progress Review', *Commonwealth of Australia,* 2014

9. WYVERN submarine, https://www.bmt.org/projects/project/3322/wyvern, (Accessed 04/2023)

10. TJIPARURO, Z., and THOMPSON, G., 'Review of Maintainability Design Principles and Their Application to Conceptual Design', *Proc. Instn Mech. Engrs Vol. 218 Part E*, 2004

7. AUTHORS' BIOGRAPHY

Ross Millard BEng, AMRINA holds the current position of Senior Naval Architect in the Submarines team at BMT. He is responsible for design reviews and third-party assurance of in-service and new build RN Submarines. His previous career experience includes the integrity management and life extension of fixed and floating offshore oil and gas assets across the world.

Alex Walchester, CEng MEng MRINA, holds the position of Senior Naval Architect at BMT. He is responsible for undertaking naval architecture work, primarily focussed on submarines. His previous experience includes tasks ranging from concept design to in-service design change. Previously he has presented on the use of reference designs and using submarines for maritime security.

EXPLORATIONS OF NOVEL NUCLEAR PROPULSION FOR SUBMARINES

R J Pawling, UCL Department of Mechanical Engineering, UK

SUMMARY

With a handful of exceptions, most notably the lead-bismuth cooled reactors in the Soviet-era ALFA fast attack submarines, all nuclear submarines to date have used a pressurised water reactor, in either dispersed or integral configurations. As far as is publicly known, all current programmes also use this technology, which is in contrast to the variety of different technologies known as "Generation IV", such as high-temperature gas cooled reactors, under development for land-based use. This paper summarises some of these possible alternatives to the PWR, illustrated with UCL student designs developed in the three-month Submarine Design and Acquisition Course. Some aspects of the Russian KANYON strategic weapon are discussed and some aspects of submarine design with novel reactors are highlighted.

1. INTRODUCTION: THE UCL SUBMARINE DESIGN AND ACQUISITION COURSE

The Naval Architecture and Marine Engineering (NAME) and Marine Research Group (MRG) are divisions of the UCL Department of Mechanical Engineering. The NAME being focussed on providing postgraduate and undergraduate education to naval architects and marine engineers, whilst MRG is a research grouping. Research and teaching interests overlap with student design projects featuring novel areas of interest.

One teaching offering of the UCL NAME is a Submarine Design and Acquisition Course (SDAC), a 9-week postgraduate course, running in the autumn or winter with four weeks of technical lectures from UCL staff and industry experts, covering all aspects of submarine design from the CONOPS to stability and structures to weapons, escape etc [1]. Students work in small groups to develop a concept design in response to a User Requirements Document (URD) provided by the UCL staff. Requirements have a unique "special feature" such as the incorporation of a new technology, an unusual restriction on the design, or a novel concept of operations. Table 1 gives some examples of recent design requirements.

Table 1. Some recent UCL SDAC designs

YEAR	NAME	SUMMARY
2019	RENEWABLE	An all-electric submarine recharging from ocean energy infrastructure, described in [2]
2019	PUNCHY	A submarine to deploy large strategic torpedoes
2021	ENVELOPER	A submarine to covertly deploy a strategic surface to air missile battery
2021	AVIATOR	A submarine to provide broad-area ASW cover using Uncrewed Air Vehicles
2022	RAIDER	A conventionally powered amphibious assault submarine

To generate propulsion packages in support of the SDAC and also surface ship designs, the author has developed several simplified scaling models for different types of nuclear reactors, based on public data from IAEA, research papers and declassified material. It should be noted that the author is not a nuclear engineer and these are sizing not design models. For example a reactor core for a Pressurised Water Reactor (PWR) is sized using a value for MWhrs per tonne heavy metal, tonnes per cubic metre and an overall aspect ratio – not a nucleonic model. Some aspects such as shielding thicknesses have been checked with a separate calculation but are assumed to be constant for the relatively small ranges of powers (and thus core sizes) of interest. These models are used to size individual components, which are then arranged in a 3D CAD model to allow overall compartment size, and thus shield area and weight, to be determined.

2. ALTERNATIVES TO THE PRESSURISED WATER REACTOR (PWR) IN SUBMARINE DESIGN

The PWR, with power conversion by the Rankine cycle – steam turbines – and mechanical propulsion via a gearbox is the most common arrangement for nuclear powered submarines. There have historically been a small number of exceptions, such as the one-off USS Seawolf (SSN-575) which was originally completed with a liquid-sodium cooled reactor, again using Rankine-cycle power conversion. SSN-575 was later converted to a use a PWR. Whilst the US Navy did investigate other types of reactors, the PWR having achieved operational status first, combined with Admiral Rickover's centralised control of the programme limited research into alternatives [3], [4]. Similarly, although the UK briefly considered a submarine derivative of the gas-cooled Calder Hall reactor [5], PWR technology was adopted with a UK design (PWR-1) containing limited US input [6] [7]. The USSR also experimented with the liquid sodium cooled reactor, before switching to a lead-bismuth eutectic coolant for the Project 705 Lira (NATO: ALFA) class, with seven boats built, each with a single OK-550 or BM-40A reactor [3] [8].

Submarine PWRs have amassed many ship-years of service and a well-understood technology, particularly in the fields of materials and chemistry and reliability, which unlike core design or thermal-hydraulics are difficult to predict by modelling. On the land side it should be noted that the organisations such the IAEA and associated Gen-IV International Forum are actively promoting a much wider range, with six novel families of interest in addition to improvements of PWR and Boiling Water Reactor (BWR) technology [9]. Just as with technologies such as energy storage, UXVs and AI, it is possible that militaries may find themselves adapting commercially developed nuclear concepts and systems, rather than being the technology leader.

3. RADIOISOTOPE THERMAL GENERATORS (RTGS)

The lowest-power option in the nuclear propulsion space is the RTG. These devices are more commonly used in space probes for the outer solar system [10] and use thermoelectric devices to convert the decay heat from small capsules of radioactive material into electricity. They are not nuclear reactors in that no chain reaction occurs; instead atoms decay naturally. RTGs have a number of limitations; they rely on plutonium as a fuel; their power output will reduce over time; they cannot be "throttled" in any way; overall efficiency is generally low due to the use of thermoelectric devices for energy conversion. However, they have no moving parts and solid-state construction makes them very robust. RTGs have seen application in the maritime sector, principally to power various types of buoys [11] or lighthouses.

4. LOW-PRESSURE PRESSURISED WATER REACTORS

Low power variants of the basic PWR concept have been proposed for marine propulsion. The AMPS (Autonomous Marine Power Source) and "Slowpoke" reactor concepts developed in Canada in the 1980s are the main example, being proposed for a modification to an existing SSK design such as the Upholder class. The research submarine SAGA was originally intended to be built with a 100kWe AMPS reactor [12][13], but this was later changed to a conventional propulsion system with the external reactor compartment adapted into a diver LI/LO chamber.

The AMPS-1000 was low pressure (3.67MPa) reactor design to produce 1.4-1.7MW (depending on seawater temperature) using TRIGA uranium-zirconium-hydride fuel. This fuel is notable for proliferation resistance and also a strongly negative thermal coefficient that makes it resistant to transients and prompt criticality, the latter allowing the reactor to be brought critical at low temperatures, with an abbreviated start-up time of 90 minutes [14]. Figure 1 illustrates the AMPS-1000 design, including the large reserve coolant tank surrounding the reactor pressure vessel. The hydrodynamic ports are a type of fluid valve that would ensure immediate make-up of the primary circuit with the reserve cooling water if the primary circuit pressure or velocity was to drop.

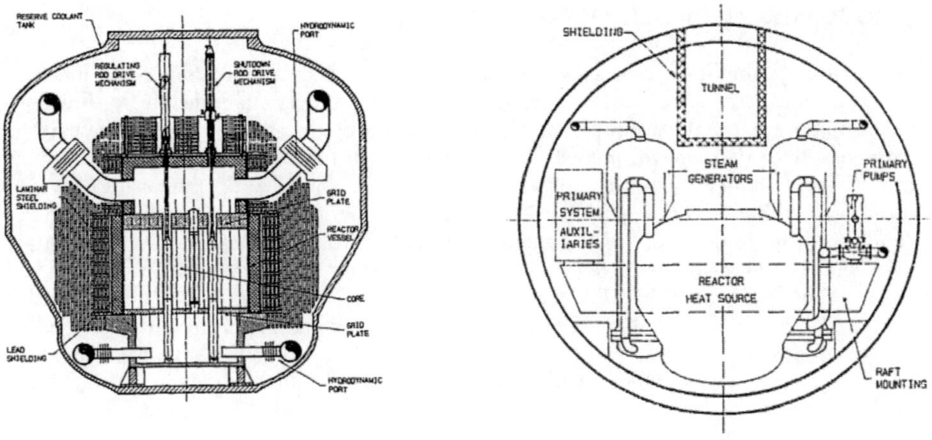

Reactor arrangement showing reserve cooling tank

Section through proposed submarine installation, 7.3m hull diameter

Figure 1: The 1.5MW AMPS-1000 low-pressure submarine reactor proposal [14]

With a coolant outlet temperature of only 207 °C (roughly 100 degrees less than a conventional PWR) and a subcritical saturated steam Rankine cycle for power conversion, the AMPS-1000 was not power-dense or highly efficient – net thermal efficiency being 15.7%. The advantage being the high level of safety including relatively large temperature margins to various accident scenarios.

4.1 BOREAL, 2018

In 2018 a student group was given the task of designing a submarine using an AMPS-1000 type reactor, notionally for the Canadian navy. The students arrived at a double hulled sailless design with all electric propulsion using contra-rotating Rim Driven Propulsors (RDPs). The relatively large size of the propulsion machinery and boost battery, within a limited overall budget, had the consequence of a modest weapons outfit considering the size of the submarine.

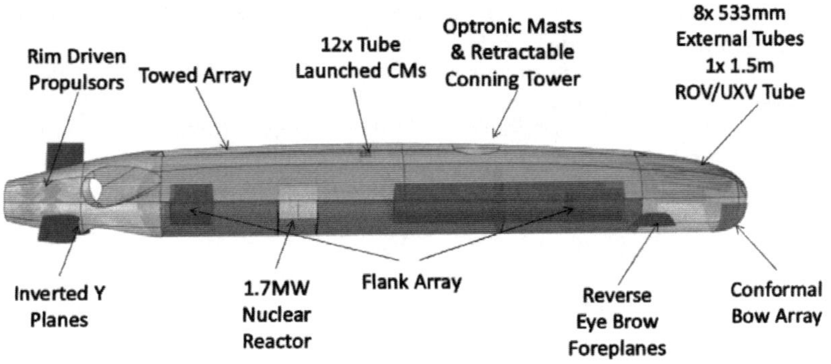

Figure 2: Side view of the BOREAL design showing the main features

Table 2. BOREAL principal particulars

Length OA	78.9m	**Sustained speed**	20kn
Length PH	44.7m	**Sprint speed**	24kn for 20 hours
PH diameter	7.85m	**Surfaced displacement**	2723te
DDD	300m	**Submerged displacement**	3284te
Tubes	8	**Weapons**	8
ROVs	3	**Countermeasures**	12
Crew	66	**Stores endurance**	80 days

5. LIQUID METAL COOLED REACTORS

Liquid metal cooling using sodium or lead-bismuth eutectics has been applied to experimental fast-breeder reactors on land – these coolants do not thermalise neutrons and so are compatible with the breeding process. Neutron moderation is possible in liquid metal cooled reactors by the addition of graphite beryllium in the reactor core, and operational naval plants have all operated in the thermal neutron spectrum. The main advantages of liquid metal cooling for a power reactor are; the ability to operate at high temperature but low pressure; compatibility with superheated steam cycles; and a substantial temperature margin before coolant boiling occurs in the event of a shutdown without external cooling available. These plants have proved problematic in service, with Seawolf being converted to a conventional PWR and the Soviet boats suffering several accidents [16] [8]. However research continues into liquid metal cooled reactors, which are one of the highlighted Gen-IV types [9]. The "Alfa" class submarines were built in two versions with different reactor designs. The BM-40A has evolved into the commercially-oriented SVBR series, specifically the 280MWth / 100 MWe SVBR-100 [17], intended for applications in floating nuclear power plants and illustrated in figure 3.

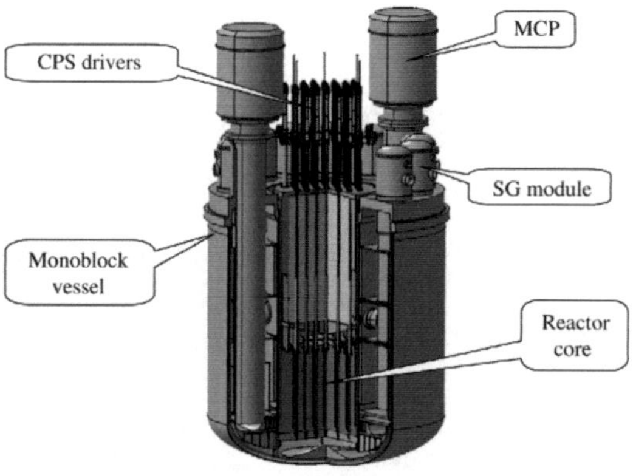

Figure 3: SVBR 100 reactor vessel with multiple steam generators [19]

In practice, the theoretical high power density is difficult to achieve – lead and lead alloys have compatibility issues with steel over 450C, and fast-moving liquid metal erodes piping significantly faster than water, limiting flow rates or requiring unusual materials for the primary circuit [18]. The modern Russian designs appear to have lower power densities than their historical predecessors, possibly due to these limitations. Note that figure 3 shows the reactor vessel and its integral steam generators (SG) and pumps (MCP) – the actual reactor compartment is much larger as steam separator drums are used to combine the SG outputs.

5.1 QUICKSILVER, 2019

This student group were provided with a 12MWs version of the SVBR-100 and required to design the a "modern ALFA" – an interceptor submarine with high speed without boost batteries. The resulting design, Quicksilver, is summarised in Table 3 and illustrated in Figure 4 with the principal particulars in table 3. To achieve high speed with modern lead cooled reactor power densities, two plants were required.

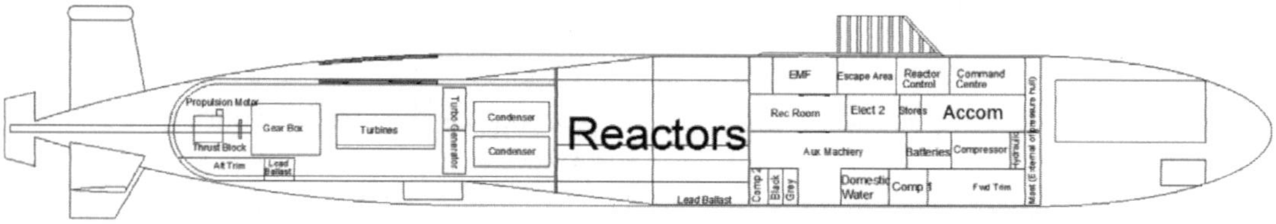

Figure 4: QUICKSILVER section view [20]

Table 3. QUICKSILVER characteristics

Length OA	84m	**Sustained speed**	42kn
Length PH	57.8m	**Submerged displacement**	4700te
PH diameter	10m	**Crew**	32+4
DDD	275m	**Stores endurance**	15 days
Tubes	10	**Weapons**	10

6. HIGH TEMPERATURE GAS COOLED REACTORS (HTGCR) AND NUCLEAR GAS TURBINES

Gas cooled reactors made up a significant part of the UK electrical generating capacity through the latter part of the 20th century, with the Magnox and Advanced Gas Reactor designs. Carbon dioxide is used to cool the reactor core, with boilers generating superheated steam, with reactor outlet temperatures around 6400C [21]. Gas cooling permits much higher operating temperatures and power conversion via Brayton cycle machines (i.e. gas turbines). The higher operating temperature permits higher thermal efficiency and power density, but makes selection of materials for the core, primary circuit and coolant material more difficult. Experimental land-based plants such as DRAGON using helium coolant operated with outlet temperatures of 750C [22]. High temperature gas cooled reactors generally have two topologies: direct and indirect. In the direct case, the primary coolant gas passes through the core, then carries out work in a turbine, before returning to the core. In the indirect case the primary circuit gas transfers heat energy to a secondary circuit gas, which passes through the turbine. The secondary circuit is usually closed-cycle, but open-cycle arrangements are possible – essentially the heat exchanger replaces the combustion chamber in a conventional gas turbine.

HTGCR have long been of interest for portable and mobile applications. The US Army built and operated the ML-1 direct cycle nitrogen cooled Brayton cycle nuclear reactor in the 1960s [23]. This 300-500kW plant could be transported on two truck trailers and was intended to power bases, allowing hydrocarbon fuels to be reserved for vehicles. The US Airforce sponsored Aircraft Nuclear Propulsion programme led to the construction of a prototype nuclear turbojet reactors; an open direct cycle plant by General Electric, in which external air passed through the reactor core, and a closed indirect cycle plant by Westinghouse [24]. This programme led to a large number of derivative designs, with the technology refined and proposed for transport applications. These included the GE type 630A "Maritime Nuclear Steam Generator" [25], and type 601B podded reactor for use in surface ships and submarines [26].

Miller [27] describes a later Westinghouse design for a helium-cooled reactor with a direct closed-cycle gas turbine including heat recovery via an intercooler. The reference plant was notable for a very high power density, designed to replace LM2500 gas turbines as shown in Figure 5.

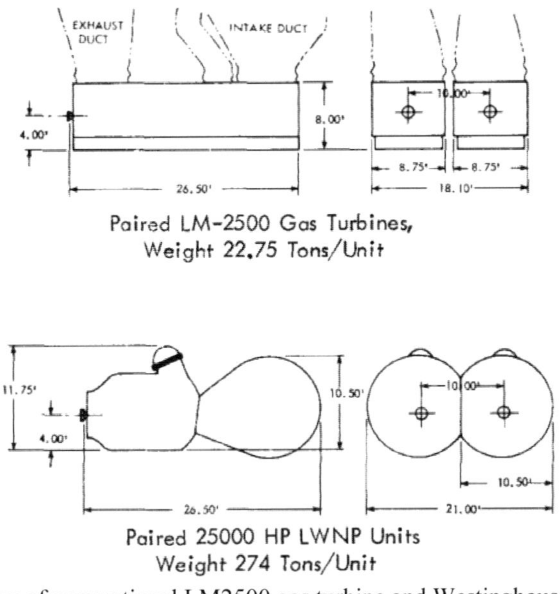

Figure 5: Comparison of conventional LM2500 gas turbine and Westinghouse nuclear plant [27].

A particular design feature of HTGCR designs is their approach to shielding. Most designs for mobile or portable plants aim for a very compact package, with power conversion machinery mounted in a single unit with the reactor core. Shielding is tightly wrapped around the core, which leads to some complications in design to properly shield the openings for cooling gas [27]. Some designs are sufficiently shielded to allow a crewed engine room. Whilst this significantly reduces total weight and space demands, it has clear implications for maintenance. The compartment housing a PWR can be entered some time after shutdown, and even reactor components such as control rod drives can be accessed. This is not the case where they are fully enclosed in a "shrink wrap" type shield.

The application of HTGCRs to submarines has been considered by several authors. Gouge [28] described the thermodynamics of direct and indirect Brayton cycles for a helium-cooled high temperature nuclear reactor. These included recuperators to achieve a high overall thermal efficiency of 44%. An additional feature of note was the use of electric propulsion, with a turbogenerator operating at 3600 rpm. The advantages of reduced plant size and weight have been noted by [28] and [29]. McDonald [30] summarised the range of technologies that would be required (or had already been demonstrated) to enable a nuclear gas turbine for submarine propulsion, as illustrated in figure 6. Considering this, it is however easy to understand why the PWR – which fit well with contemporary understanding of steam systems – was adopted in the mid 20th century.

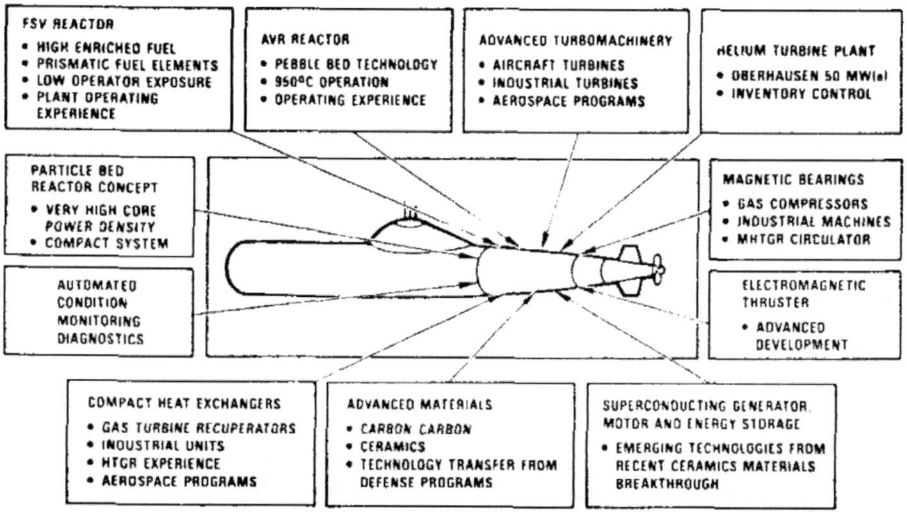

Figure 6: Enabling technologies for a submarine nuclear gas turbine propulsion system [30]

6.1 FILIAL, 2012

A scaled-down derivative of the Westinghouse reactor was provided to one design group in the 2012 UCL Submarine course, with the requirement to design a patrol submarine with long transit range. The resulting design, named FILIAL as the principal particulars summarised in Table 4. The students adopted a novel propulsion arrangement: the 4MW HTGCR would provide transit power in an all-electric arrangement, being shut down for patrol. On patrol at quiet speed (4 knots), power would be provided by PEM fuel cells running on hydrogen and oxygen, with boost power for sprint coming from lithium-ion batteries [31]. The logic behind this arrangement was that the energy requirements for transit on fuel cell power alone would lead to huge reactant tanks, whilst the fuel cell was much quieter than the reactor for patrol.

Table 4. Principal particulars for the 2012 FILIAL design [31]

Length OA	73m	**Sustained speed**	16kn
Length PH	38m	**Sprint speed**	20kn
PH diameter	9m	**Surfaced displacement**	2904te
DDD	370m	**Submerged displacement**	3200te
Tubes	4	**Weapons**	18
VLS	14 (2 x VPM)	**Countermeasures**	40
Crew	50	**Stores endurance**	90 days

7. HEAT PIPE REACTORS

A heat pipe is a type of passive heat transmission device that uses passive capillary forces rather than convection, pumps etc. They are commonly used in laptop computers, where the desire for a very thin package means that cooling fans or heat sinks cannot be located over the processors. The basic principle of the heat pipe is shown in figure 7. Whilst the heat carrying capacity of an individual heat pipe is limited, when operating at low load (for example in a post-shutdown state) they are robust against changes in orientation [32] and can even be bent, a solution used in designs for space application to route the heat pipes around a shadow shield.

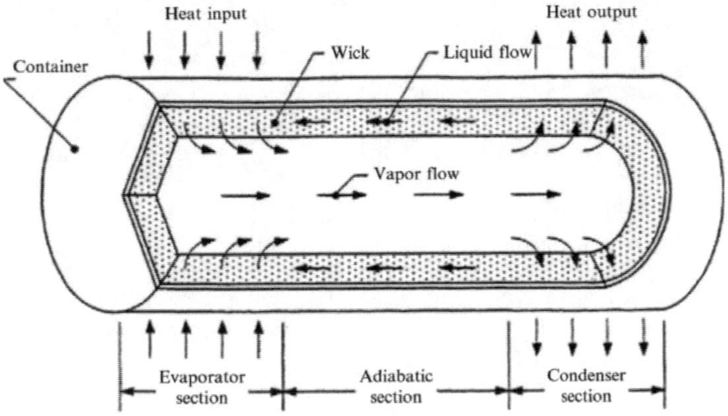

Figure 7: Concept of operation of a heat pipe [33]

Heat pipe reactors have been proposed for use in space applications [34] [35] and more recently for portable terrestrial use, in both power and combined heat and power applications, e.g. [36] [37]. These generally feature a horizontal reactor core with heat pipes projecting from one or both ends as illustrated in figure 8.

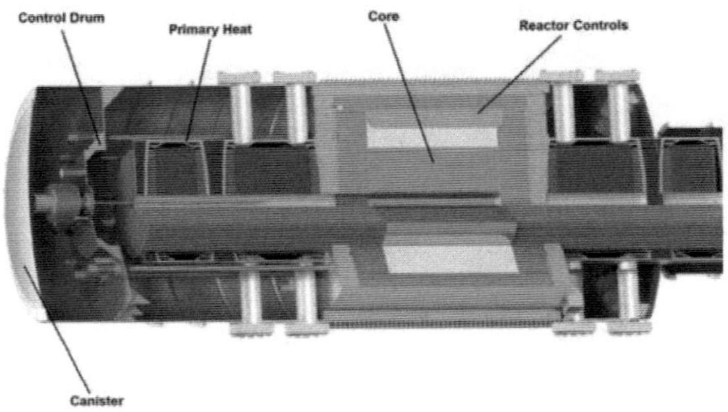

Westinghouse eVinci [36]

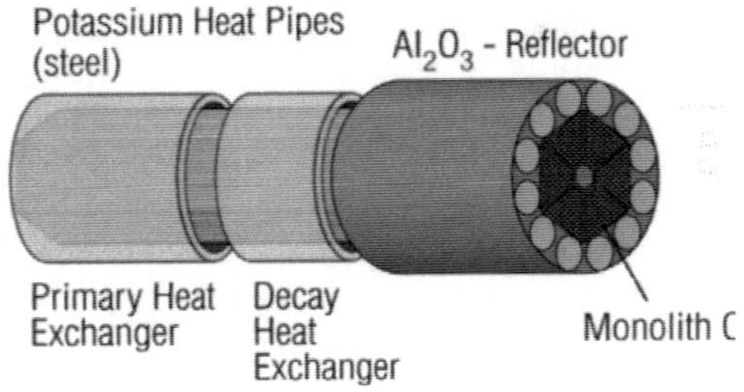

LANL Special Purpose Reactor [37]

Figure 8: Examples of recent developments in terrestrial heat pipe reactor designs

These designs have a common topology of a core monolith consisting of solid material with holes machined for fuel assemblies and heat pipes. Load following and negative thermal coefficient is achieved through phenomenon such as thermal expansion of the core and doppler broadening. Being small cores, neutron economy is very important and a thick neutron reflector (typically tungsten) is required. This also forms the control and shutdown method, with the reflector containing control drums as shown in Figure 9 below. The reactor is brought critical by rotating the control drums so that the reflector faces inwards. This technology has been used operationally in space reactors such as TOPAZ.

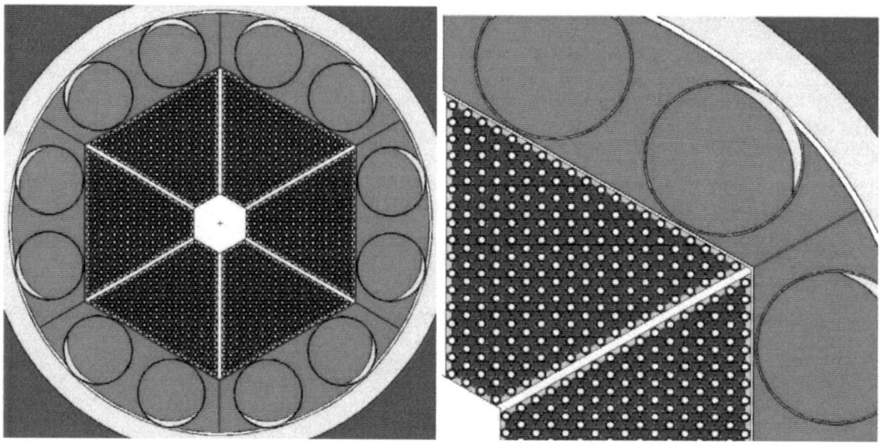

Figure 9: Section of heat pipe reactor core, showing multi-segment core monolith, neutron reflector with control drums featuring a crescent boron carbide neutron absorber. Reactor is critical in this image. [38]

With operating temperatures in the hundreds of degrees, heat pipe reactors can be applied to several possible energy conversion processes, including thermoelectric, Brayton [39] or Stirling [40] and for land applications may also provide heat energy for process and space heating.

7.1 HEAT PIPE REACTORS IN UCL SUBMARINE DESIGNS: DEPLOY (2017)

UCL has used heat-pipe reactors in several submarine designs. The first was in 2017 when group DEPLOY were assigned the requirement to design a medium to high capability SSKn using a notional 190kWe heatpipe reactor with thermoelectric power conversion. In this early concept, the entire reactor and power conversion system was integrated into a compact shielding cylinder, to be replaced as a complete unit. In an example of how students will not always do what one expects, group DEPLOY noted the requirements did not specify *how many* reactors to use, and so fitted five, as can be seen in figure 10.

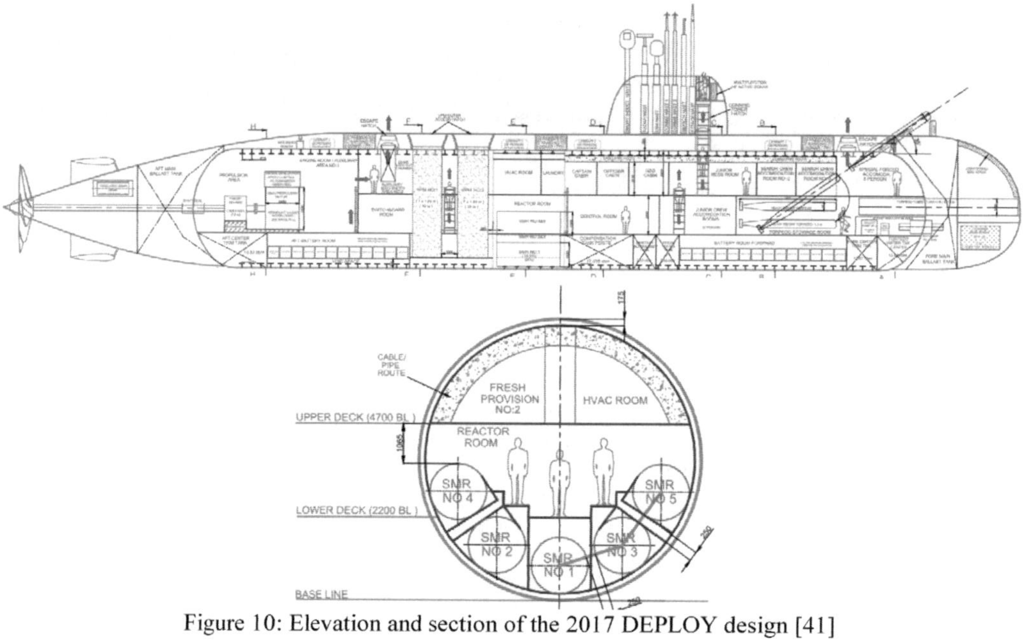

Figure 10: Elevation and section of the 2017 DEPLOY design [41]

Table 5. Principal particulars of the DEPLOY design

Length OA	64m	**Sustained speed**	12 knots
Length PH	46m	**Sprint speed**	23 knots
PH diameter	7.5m	**Surfaced displacement**	1882te
DDD	350m	**Submerged displacement**	2322te
Tubes	4	**Weapons**	12
VLS	8	**Countermeasures**	Yes
Crew	44	**Stores endurance**	100 days

7.2 CRAWLER, 2018

Team CRAWLER were given a similar requirement; an SS(n) with long endurance and high ASW capability, but powered by a single "nuclear AIP", in this case a second iteration of the UCL-sized Heat Pipe Thermoelectric Reactor (HPTER), rated at 160kWe but with designed so that the core and heat pipe module can be removed separately from the energy conversion units (to allow refuelling and maintenance in situ). Table 6 summarises the principal particulars of the CRAWLER design. With extremely limited power available from the micro-reactor, the students elected to use diesel power to deploy whilst snorkelling, with the power for a slow speed patrol provided by the reactor. This design required much effort from the students in accurately calculating hotel loads, as most UCL scaling data is for larger SSNs and SSKs.

Table 6. CRAWLER design principal particulars [42]

Length OA	58m	**Patrol speed**	4 knots
PH diameter	6m	**Sustained speed**	10 knots
DDD	200	**Sprint speed**	18 knots
Tubes	4	**Surfaced displacement**	1581te
Weapons	10	**Submerged displacement**	1700te
Crew	24+6	**Stores endurance**	60 days

7.3 NAVIGATOR, 2022

The third iteration of the UCL heat pipe reactor design was introduced in 2022, after sighting early press material regarding the latest research developments by Rolls Royce, which appeared to show gas-turbine derived Brayton cycle power conversion machinery. A notional reactor was sized and is illustrated in Figure 11, along with the heat exchanger units. Eight micro turbines driving high speed alternators were adopted, primarily to reduce overall package length and weight, but also to provide redundancy. With a core life of approximately five years (depending on operational profiles), access for refuelling as a unit was explicitly required.

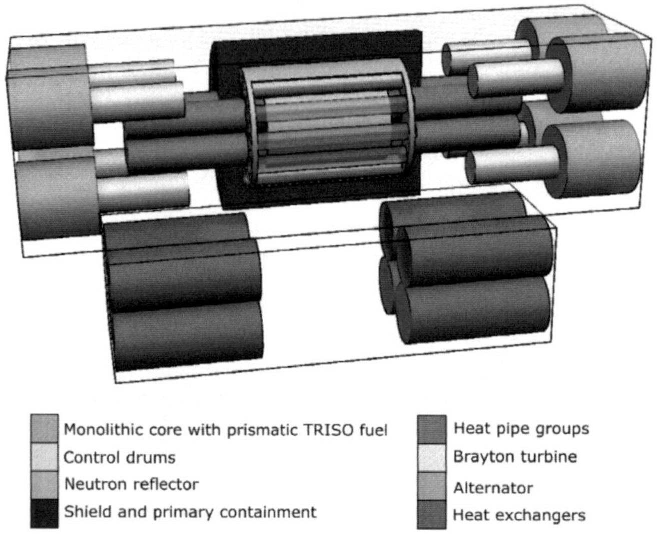

Figure 11: The notional UCL 5MWe Heat Pipe Gas Turbine Reactor (HPGTR)

With dimensions a little smaller than a 40 foot container, the reactor represented an unusual arrangement problem, in that it is quite difficult to arrange a *long* item in a submarine, as opposed to a *tall* one. The students elected to fit the reactor in its' own pressure hull, between the main hull and Rim Driven Propulsor (RDP) aft. This is illustrated in Figure 12.

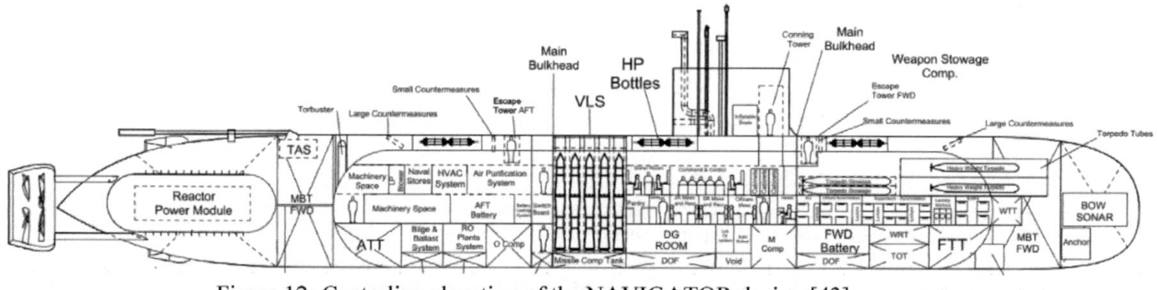

Figure 12: Centreline elevation of the NAVIGATOR design [43]

This arrangement had the added benefit that the heat exchangers could potentially be simplified as the pressure hull and surrounding free-flood space now acted as the heat sink, with heat transfer improved by the external framing on the smaller hull. The concept of a separate hull for the reactor has appeared in historical designs, such as the VAU-6 "auxiliary reactor" fitted to a single Soviet Project 651 JULIETT class for trials, or the postulated arrangement of the LOSHARIK special purpose submarine [41]. The principal particulars of NAVIGATOR are summarised in table 7 below.

Table 7. NAVIGATOR principal particulars

Length OA	76.7m	**Sustained speed**	20kn
Length PH	44.7m	**Sprint speed**	26kn
PH diameter	8m	**Surfaced displacement**	2561te
DDD	300m	**Submerged displacement**	2826te
Tubes	4	**Weapons**	12
VLS	10	**Countermeasures**	14
Crew	40+8	**Stores endurance**	90 days

8. BATTERY BOOST

Whilst the exact capabilities and sizes of the various novel reactor designs developed at UCL have varied, there have been some common aspects. Most notably, all use a battery system for sprint, with the reactor only capable of providing a transit or patrol speed. The BOREAL, DEPLOY and NAVIGATOR designs can self-deploy on reactor power, whilst CRAWLER deploys under diesel power whilst snorting, with the very low power reactor only used for the patrol speed of 4 knots. Table 8 below summarises the speed and endurance capabilities of the hybrid reactor-battery systems. Notable is BOREALs very large battery capacity, which was selected to allow the boat to escape from under the arctic ice cap in the event of a reactor failure.

Table 8. Performance characteristics of UCL nuclear-battery hybrid submarines

	BOREAL	CRAWLER	DEPLOY	FILIAL	NAVIGATOR
Reactor Speed	14	4	12	16	20
Sprint Speed	24	18	23	20	26
Sprint Endurance	20 hours	2.5 hours	2.17 hours (50nm)	-	2 hours
Energy Storage	107.5MWhr	4.1MWhr	7.9MWhr	7.5MWhr	9MWhr

9. A REPRESENTATIVE 500-TONNE NUCLEAR GAS TURBINE SUBMARINE

"Project Nobska" was a study commissioned by the US Navy in 1956 to investigate the impact of advanced technology on undersea warfare which concluded by promoting several design directions and technologies, such as an increased focus on submarine performance (as seen in the subsequent Permit class SSNs) and fast homing torpedoes [44]. One intriguing option not pursued is the concept of a 500-tonne submarine, intended to be built in large numbers. In addition to the idea of increasing numbers of boats by reducing the tonnes per boat, this concept is interesting in that many navies are acquiring submarines that use liquid oxygen, the resupply of which would seem an obvious target for asymmetric attacks using drones or anti-materiel rifles. A small nuclear plant might be attractive in reducing refuelling vulnerability. In this section an indicative concept design developed by the author is described.

Initial estimates showed that a 500-tonne submarine would require around 4MW to make 30 knots, and so the 5MWe heat pipe nuclear gas turbine was selected. The closest post-war analogue for a such a small submarine was seen to be the German Type 205/206 classes, which are coastal submarines. A similar weapons fit was selected, but with a large flank array, and the crew size was set at 21 based on German and Scandinavian submarine practice. The principal particulars of the resulting design are shown in Table 9, with the design illustrated in Figure 13.

The requirement to keep size small drove many design decisions. It led to a small reserve of buoyancy, due to limited volume in the ends of the hull envelope. This is however a relatively straightforward trade-off between more internal volume for slightly less speed. External torpedo tubes were adopted, with reloading via a modular "saddle" over the reduced diameter forward pressure hull. Syntactic foam blocks were fitted forward to compensate for the weight-buoyancy mismatch this created.

Table 9. Principal particulars of the indicative submarine design

Length OA	49.6m	**Sustained speed**	29kn
Length PH	41.4m	**Surfaced displacement**	665te
PH diameter	4.8m	**Submerged displacement**	705te
DDD	300m	**Weapons**	8
Tubes	8	**Countermeasures**	16
Crew	21	**Stores endurance**	30 days

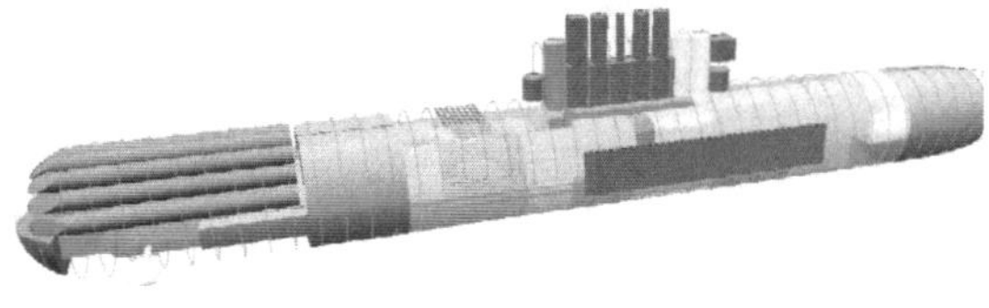

All functional groups, external hull hidden

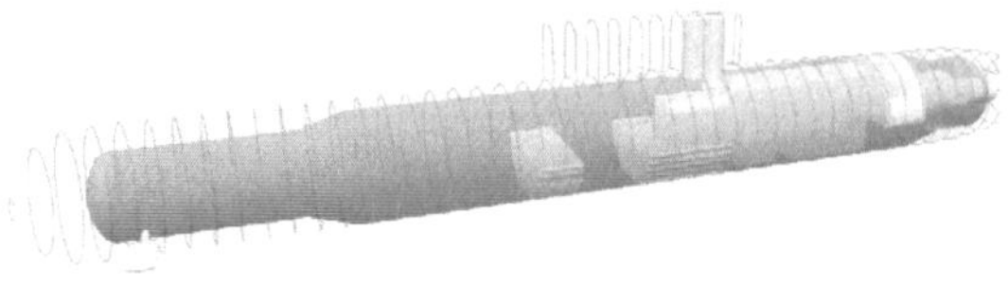

MOVE function, with pressure hull visible

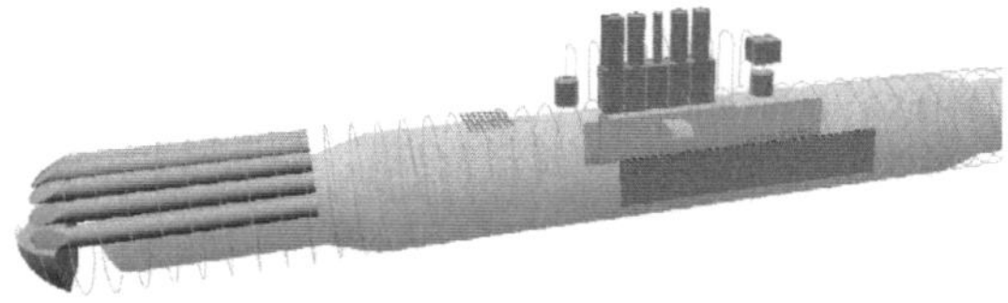

FIGHT function

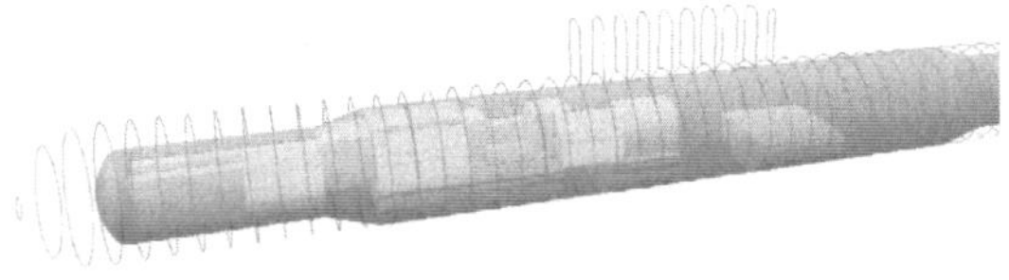

INFRASTRUCTURE function, including accommodation and auxiliaries

Figure 13: Functional groups for the example 500-tonne submarine

Whilst this study indicated that a very small submarine would be possible with the novel reactor technology, the 500 tonne target displacement was exceeded. One reason for this was the use of modern accommodation standards – although cramped by the standards of a full-size SSN, individual berths were based on US Navy surface ship equipment [45]. Associated with this was the space and weight impact of using SSN atmosphere control (e.g. electrolysers) rather than the chemical systems used on historical designs – the latter can be fitted in odd places, rather than being large units, which increased the size of the auxiliary machinery spaces. This led to the unusual pressure hull configuration with a "one and three quarter" deck amidships, to provide sufficient volume.

As with the NAVIGATOR student design, the awkward shape of the reactor module drove the design. Whilst it has much greater energy density than a diesel-electric plant, it is a single unit and cannot be fitted into convenient spaces as diesel or battery tanks can. Figure 14 highlights the propulsion machinery, including how the cuboid reactor led to an inefficient use of space. The location of the reactor relatively high in the submarine also impacted the stability – although the boat is stable both surfaced (GMtf = 0.19m) and submerged (BG = 0.228m).

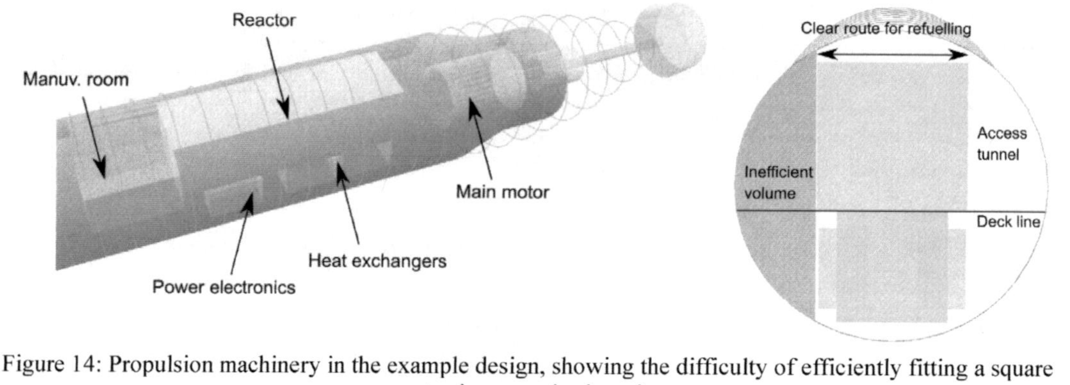

Figure 14: Propulsion machinery in the example design, showing the difficulty of efficiently fitting a square reactor in a round submarine

10. OVERALL COMPARISON OF NOVEL REACTORS

Table 10 summarises several characteristics of the current set of notional UCL reactor packages for submarines, including the four conventional PWR options. Two options are provided for the 5MWe heat pipe reactor, as there remains some uncertainty regarding the shielding design, with the "high" option having additional shielding around the cuboid "compartment".

Table 10. UCL notional reactor packages

			UCL PWR			LMCR	AMPS	HPTER	HTGCR	HPGTR	HPGTR (High)
Output Power	MW	5	10	15	20	12	1.7	0.16	4	5	5
RC Height	m							2.5		2.3	2.3
RC Dia/Width	m	7.15	8.6	9.6	10.5	7.4	5.2	4.0	2.85	2.3	2.3
RC Length	m	5.3	6.0	6.5	6.6	6.5	4.0	5.5	5.3	8.0	8.0
RC Volume	m3	213	349	470	571	280	85	55	34	42	42
RC Mass	te	391	572	733	868	385	184	72	153	80	133
Thermal efficiency		26	25	25	25	26	16	15	~40	39	39
Secondary	m3	631	814	944	1049	525	425	0.8	0	27	27
	te	222	270	304	330	165	163	0.6	0	18	18
RC	Vol/MW	42.6	34.9	31.4	28.6	23.3	50.0	343.8	8.5	8.5	8.5
	Mass/MW	78.2	57.2	48.9	43.4	32.1	108.2	450.0	38.3	16.0	26.6
Total	Vol/MW	169	116	94	81	67.0	300	348.9	8.5	13.9	13.9
	Mass/MW	123	84	69	60	45.8	204	453.6	38.3	19.6	30.2

Whilst there is some uncertainty in the values for secondary plant size – some elements such as gearboxes, motors etc. will of course be common to all types – it is clear that some of the novel options such as high temperature reactors can achieve much more compact and lighter plants. However, it is interesting to note that their advantage is reduced as installed power increases. This is due to the reduction in surface area to volume ratio for larger PWR reactor compartments, and the associated reduction in shielding mass.

11. POSSIBILITIES FOR POSEIDON

Poseidon, (NATO: KANYON) is a reportedly nuclear powered, nuclear armed torpedo developed by Russia, with a maximum speed variously reported as 70 or 100 knots [46]. This is not an entirely new idea, the Project 627 "Kit" (NATO: NOVEMBER) class SSNs were originally conceived as carriers for a single giant nuclear torpedo for attacking ports etc. As Table 9 illustrates, most nuclear power plants are very low power density, and whilst radiation shielding could be reduced for an uncrewed vehicle (although not eliminated, particularly with regards to the nuclear warhead), there remains the question of available volume. Figure 15 shows the power-speed curve for a notional KANYON of length 24m, diameter 2m and volume of displacement 68m3.

Although Russian submarine PWRs are generally more compact than their western counterparts, due to differences in steam generator and piping design, and access for maintenance would not be required in a torpedo, examining the mass values in Table 9, the weight of the power conversion machinery (turbines and condensers) would appear to rule out any Rankine cycle machinery, even a superheated steam plant, as would be possible with a liquid metal cooled reactor.

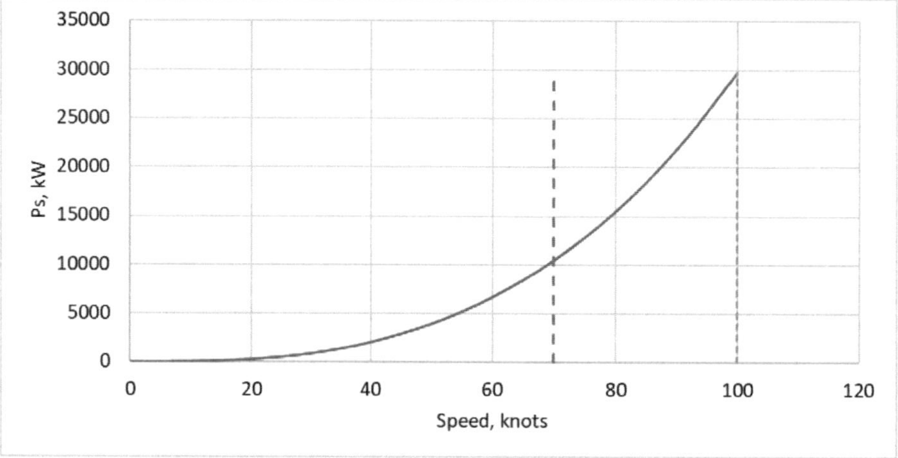

Figure 15: Power-speed curve for a notional KANYON vehicle

A nuclear gas turbine (Brayton cycle) would appear to offer the required power density, and scaling from the design described by Miller, [27] for a shaft power of 11MW, but omitting shielding, leads to a package size of approximately 5m length and 2m diameter – the latter dominated by the power conversion system. The weight of this package would be approximately 18 tonnes. However, these values exclude the cooling systems required – assuming an overall efficiency of 40%, typical of designs for closed Brayton cycles, at maximum speed some 27.5MW of heat would need to be removed by some combination of precoolers and intercoolers. This represents a significant challenge for heat exchanger design due to the short dwell time of cooling water and brings us to the main question regarding KANYONs propulsion system: where are the cooling water inlets?

Figure 16: Composite image showing the KANYON weapon [46]

Figure 16 is a composite image produced from Russian MoD by Sutton [46] and shows a very clean exterior for the KANYON torpedo, with no obvious inlets or outlets for a cooling system – of course they could be out of sight or this may be a mockup. An interesting possibility is the use of a heat exchanger wrapped around the entire hull, as was suggested for torpedoes in US Patent 3109401 [47] and also proposed by General Electric for their Type 601B closed-cycle nuclear gas turbine propulsion "pod", as shown in Figure 17. Such a configuration of heat exchanger could be hidden behind panels jettisoned after weapon launch and would be attractive considering the overall configuration, but requires further thermofluids analysis.

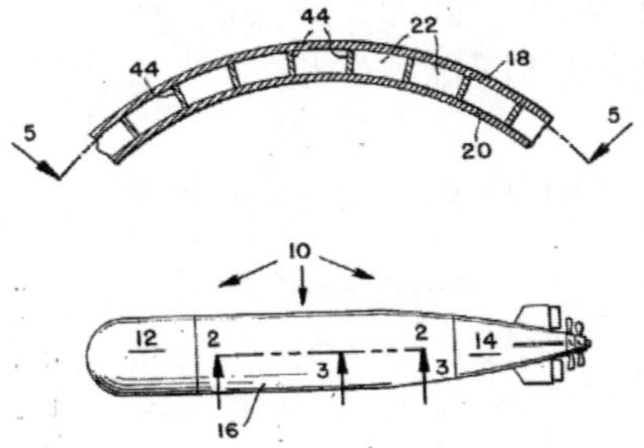

Annular heat exchanger for a torpedo [48]

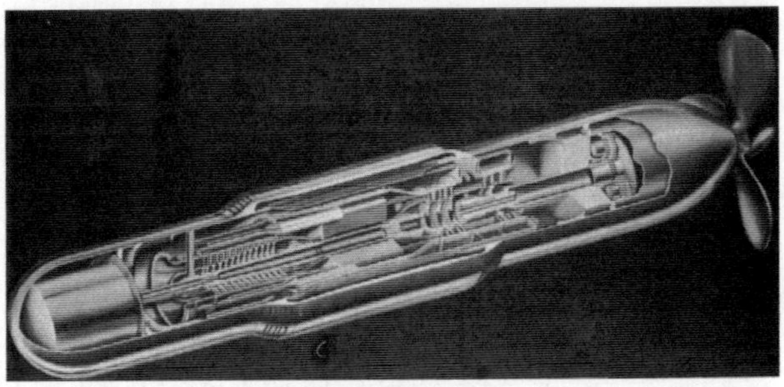

Podded nuclear propulsion system showing annular heat exchanger [26]

Figure 17: Proposed annular heat exchangers for closed-cycle propulsion systems

12. SUMMARY AND CONCLUSIONS

Whilst the Pressurised Water Reactor has been the most common type of propulsion for nuclear powered submarines, enjoying a good safety record and a high level of institutional engineering knowledge and confidence, it is not the only technology that has been considered or applied to submarine propulsion. This paper has reviewed several different alternative reactor types that have been proposed for use in submarines, and which are of interest in the civil world for use on land or in space, some of which are now experiencing funding for development.

Several iterations of the UCL Submarine Design and Acquisition Course have seen student groups assigned the requirement to use indicative reactor packages representing these technologies developed by UCL from published data. Of particular interest has been the application of nuclear power to smaller submarines, and the alternative technologies seem viable, particularly when combined with battery boost. The various UCL studies and parametric models have also indicated that a smaller reactor is not always more convenient to arrange, particularly when unitised into module containing the reactor and the power conversion system, and when short core life leads to a requirement to frequently refuel the reactor.

This is one aspect of a dynamic in nuclear submarine design where the reactor may be the driver of the design (by setting hullform diameter), but reducing the reactor size does not necessarily make the submarine smaller, due to aspects including; layout, combat system capability, accommodation and equipment arrangements, sprint capability. An unanswered question is if there can be found an "ideal" size of reactor for a given set of requirements.

Based on the history of reactor design and experimentation in the 1960s-1970s, a possible answer to the question of technical viability for the KANYON nuclear torpedo has been derived, in the form of a high temperature reactor with a Brayton cycle power conversion system, possibly using a novel heat exchanger arrangement, with the latter aspect highlighted as a feature to look out for in any images of the weapon.

This paper does not set out to criticize the use of PWR technology for submarines, rather it explores some of the history of alternative technologies and some example designs. As civil developments push the technologies, these may become more attractive as a naval option, or important to understand due to adoption by adversaries.

13. ACKNOWLEDGEMENTS

The author would like to acknowledge the hard work of the students on the UCL SDAC whose designs are described in this paper.

14. REFERENCES

1. https://www.ucl.ac.uk/mechanical-engineering/study/other-programmes/submarine-design-acquisition-course

2. PAWLING, R., DOS SANTOS, J., YIGIT OGUZ, M., WEATHERALL, D., and LANGO, B., (2020) "RENEWABLE: An Innovative UCL Concept Submarine Design for an All-electric Boat", *RINA Warship 2020*, 17 – 18 June 2020, Bristol, UK

3. POLMAR, N., and MOORE, K J., (2004), "Cold War Submarines", *Potomac Books Inc.*

4. COWAN, R., (1990) "Nuclear Power Reactors: A Study in Technological Lock-in", *The Journal of Economic History, Volume 50, Issue 3*, September 1990, pp. 541 – 567

5. HILL, C N., (2013), "An Atomic Empire", *Imperial College Press*

6. HILL, R., (2005), "Admiral Hyman G Rickover USN and the UK Nuclear Submarine Propulsion Programme", *The Thomas Gray Memorial Lecture, IMechE*, April 2005

7. MICHAEL, J G., (2019), "The Development of Nuclear Propulsion in the Royal Navy, 1946-1975", *University of Plymouth*

8. ZRODNIKOV, A V., et al, (2000), "Use of Russian Technology of Ship Reactors with Lead-Bismuth Coolant in Nuclear Power", https://inis.iaea.org/search/search.aspx?orig_q=RN:31058476

9. https://www.gen-4.org/gif/

10. SCHMIDT, G., and SUTLIFF, T., (2008), "Radioisotope Power: A Key Technology for Deep Space Explanation", *IAC* 2008

11. MORRISON, J H., (1969), "Radioisotope Power Sources for Buoy-Type Applications", *Oceanology International 69.*

12. HEWITT, J S., WILKINS, P., and KASTNER, G A., (1986), "SAGA-N: A Nuclear-Powered Submarine Vehicle for Commercial Operation", *CNS 6th Annual Conference*, 1986

13. MACDONALD, A G., and KASTNER, G A., (1985), "AMPS – A Small Nuclear Reactor Based Autonomous Marine Power Source for Sub-Sea Vehicle Applications", *4th International Symposium on Unmanned Untethered Submersible Technology*

14. OLIVA, A F., (1989), "The AMPS 1000: An Advanced Reactor Design for Marine Propulsion", *CNS 10th Annual Conference*, 1989

15. PATTISON, R., WHITEHEAD, S., LAU, J., and AIKEN, S., (2018), BOREAL UCL SDAC design report

16. LOEWEN, E P., (2012), "The USS Seawolf Sodium-Cooled Reactor Submarine", *Address to Washington DC ANS Local Section,* 2012

17. ZRODNIKOV, A V., et al, (2001) "Multipurposed Small Fast Reactor SVBR-75/100", https://inis.iaea.org/search/search.aspx?orig_q=RN:32021986

18. AGOSTINI, P., et al (2013), "Pure Lead as coolant of GEN IV Fast Reactors: specific issues and comparisons with other coolants", *Seminar on "Fast Reactor Coolants", Paris*, February 2013

19. ZRODNIKOV, A V., et al, (2008), "Innovative nuclear technology based on modular multi-purpose lead-bismuth cooled fast reactors", *Progress in Nuclear Energy 50, pp 170-178*

20. GREEN, J., WARREN, S., WILLIAMSON, A., and HOON, JS., (2018*), QUICKSILVER UCL SDAC design report*

21. NONBØL, E., (1996), "Description of the Advanced Gas Cooled Type of Reactor (AGR)", *Report NKS/RAK2(96)TR-C2, Risø National Laboratory, Denmark*

22. SIMON, R A., and CAPP, P D., (2002) "Operating Experience With The Dragon High Temperature Reactor Experiment", https://inis.iaea.org/search/search.aspx?orig_q=RN:33033056

23. "The ML-1 Design Report", Report No- IDO-28550, *The Military Projects Division, Aerojet-General Nucleonics*, May 1960

24. STOFFEL, J., (2000), "Dreams of Nuclear Flight: The NEPA and ANP programs"

25. "Summary Report 630A Maritime Nuclear Steam Generator Scoping Study", *report GEMP-107, Flight Propulsion Laboratory Department, General Electric*, March 1962

26. COMASSAR, S., (1962), "Aircraft Nuclear Propulsion Application Studies", *General Electric*, 1962

27. MILLER, R T., "Maritime Applications of an Advanced Gas-Cooled Reactor Propulsion System", *SNAME Spring Meeting*, 1977

28. GOUGE, M J., "Closed Cycle Gas Turbine Nuclear Power Plant for Submarine Propulsion", *Naval Engineers Journal*, November 1995

29. ADAMS, R M., (1994), "Submarine Engines for the Future", *USNI Proceedings*, June 1994

30. MCDONALD, C F., (1988), "Glosed-Cycle Gas Turbine potential lor Submarine Propulsion", *Trans. SNAME*

31. HARRISON, S., KAHN, I., and PATTEN, D., (2012), "SSKN Hybrid Submarine – Filial 1 Class", *UCL SDAC design report*

32. "How Wicks and Orientation Affect Heat Pipe Performance", *Qpedia*, August 2019

33. ZOHURI, B., (2016) "Chapter 1 Basic Principles of Heat Pipes and History", *in Heat Pipe Design and Technology, Springer International*

34. BUSHMAN, A., et al (2004), "The Martian Surface Reactor: An Advanced Nuclear Power Station for Manned Extraterrestrial Exploration", *Report MIT-NSA-TR-003*, December 2004

35. TRUSCELLO, V C., (1983), "SP-100, The US Space Nuclear Reactor Power Program", *NASA JPL*, November 1983

36. ARAFAT, Y., and VAN WYK, J., (2019), "eVinvi Micro Reactor", *Nuclear Plant Journal*, March-April 2019

37. STERBENTZ, J W., WENER, J E., MCKELLAR, M G., HUMMEL, A J., KENNEDY, J C., WRIGHT, R N., and BIERSDORF, J M., "Special Purpose Nuclear Reactor (5MW) for Reliable Power at Remote Sites Assessment Report", *Idaho National Laboratory*, April 2017

38. MCCLURE., et al, "Design of Megawatt Power Level Heat Pipe Reactors", *Report LA-UR-15-28840, LANL*, 2015

39. ANANTH, K P., et al, (2017) "Portable Special Purpose Nuclear Reactor (2 MW) for Remote Operating Bases and Microgrids", 2017 *Joint Service Power Expo*, 1-4 May, *Virginia, USA*

40. POSTON, D I., MCCLURE, P R., DIXON, D D., GIBSON, M A., and MASON, L., (2014) 'Experimental Demonstration of a Heat Pipe–Stirling Engine Nuclear Reactor', *Nuclear Technology*, 188:3, 229-237

41. BRASTED, M., HOPPS, J., AHMED, K., and SARGIN, E., (2017) 'Deploy UCL SDAC' *design report*

42. HODGE, D., WREN, S., LOPEZ, S., and NEWELL, A., (2018), 'Crawler UCL SDAC' *design report*

43. SAGAN, I., NUNES, M., MATTINSON, J., and TRYBIS, L., (2022), 'Navigator UCL SDAC' *design report*

44. FRIEDMAN, N., (1994). 'U.S. Submarines Since 1945: An Illustrated Design History'. *Annapolis, Maryland: Naval Institute Press*, as quoted in https://en.wikipedia.org/wiki/Project_Nobska

45. http://www.jdasales.com/furniture/berths/

46. SUTTON, H I., (2019a) http://www.hisutton.com/Poseidon_Torpedo.html

47. KARIG, H E., (1963), "Closed Cycle Torpedo Power Plant*", US Patent 3109401*, November 5th 1963

15. AUTHORS BIOGRAPHY

Dr. Rachel Pawling is Programme Director for the MSc Marine Engineering course and a Lecturer in Ship Design at UCL, teaching the subject of ship design to undergraduate and postgraduate students. She obtained her PhD in computer aided ship design in 2007 and has subsequently continued her research in projects funded by the EC, UK and US governments.

PATROLLING THE METAVERSE

Dr R Perez Fernández, and **M Toman,** SIEMENS DI SW, Spain

SUMMARY

In this paper we will submerse in the latest paradigm of Metaverse apply to the design and fabrication of submarines, we will analyse the closed loop connections between the digital and real worlds across the value chain enable actionable insights and information decisions. The objective of this paper will be to study how to produce an optimum product, how to simulate real production priorities and constructions process, commissioning, and operations applied to submarines. We will review important aspects of the Industrial Metaverse like virtual worlds, live interactions, real world relevance, interoperability, functional economy, secure and evolution of the communications, training, Augmented Reality of the digital twin in both worlds, meeting, remote control, collaboration, continue and interactively evaluation, monitoring analysis and manage of real and virtual data.

NOMENCLATURE

AI	Artificial Intelligence
AR	Augmented Reality
CAD	Computer Aided Design
CAE	Computer Aided Engineering
CAM	Computer Aided Manufacturing
CMRE	Centre for Maritime Research and Experimentation
CNC	Computer Numerical Control
CPU	Central Processing Units
IoT	Internet of Things
NATO	North Atlantic Treaty Organization
USD	Universal Scene Description
VR	Virtual Reality
WiFi	Wireless Fidelity
xDT	eXecutable Digital Twin

1. INTRODUCTION

"The Meta-Verse; And How It Will Revolutionize Everything" from Matthew Ball [1], technology expert, thought leader and best-selling author on the Metaverse, it is no less than a new plane of existence, one that will change the way we live forever.

The current focus when it comes to the Metaverse is mostly on the world of games, online retail, and social media. Interestingly, the most significant and biggest potential of the Metaverse is within the industries: manufacturing, buildings, grid and infrastructure operators, the transportation sector and very especially Defence, in this paper we will take a tour about all these new ideas and try to focalize them in the case of Submarines.

2. THE "INDUSTRIAL METAVERSE"

The Industrial Metaverse is a world, which is always on. Where real machines and factories, buildings and cities, grids, transportation, and defence systems are mirrored in the virtual world. Why is this so important?

In a digital environment, problems can be found, analysed, and fixed quickly - or better yet, discovered before they arise. And all of this will happen in a world enabling a whole new level of collaboration. Where people can break the barriers of distance and work together across countries and continents as if they were together in the very same room, in front of the very same machines or objects. Where everyone can try out new ideas easily and quickly - and where innovation takes off.

The Metaverse is where *Virtual Reality* (VR) supports people who are working hands on, on site. A virtual realm where we can travel into the past and even into the future to understand problems and processes better and combining the real and digital worlds [5].

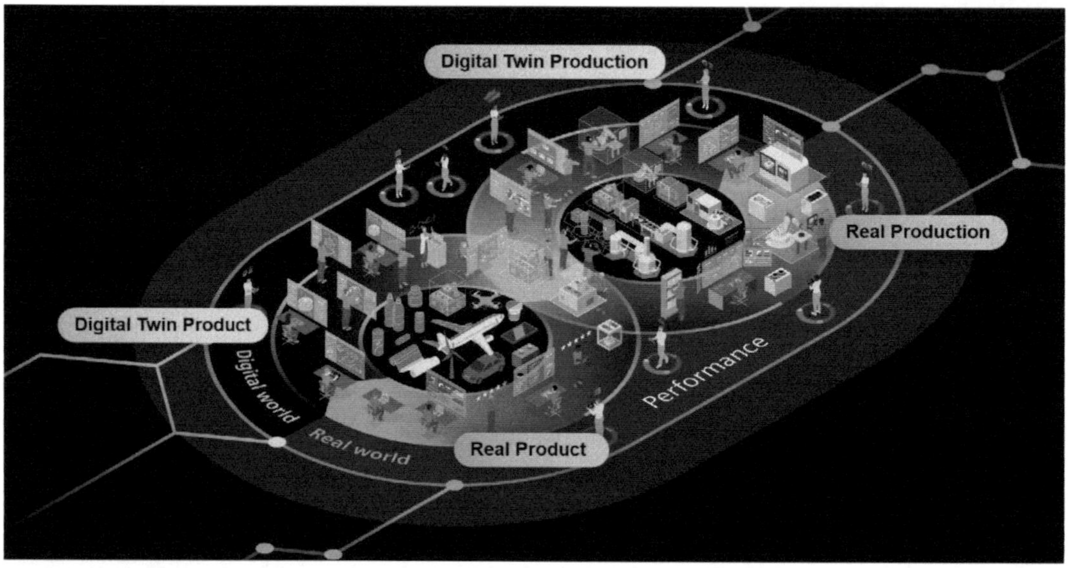

Figure 1: An interpretation on how to combine the Digital Twin concept and a Real Product.

Digital Twins already combine the real and the digital worlds to quicken the planning of factories, buildings, and entire city districts. And by connecting the digital twins to their real-world equivalents, by leveraging their data, we can improve operations over the entire life cycle.

Siemens´ Digital Native Factory in Nanjing is one example. From the very beginning we planned this facility using digital technologies. The entire factory was simulated with a digital twin, which optimized the building, and detected and mitigated potential problems at an early stage. Planning errors, both small and big, which in the past would cost a lot of money and a lot of time, were completely avoided. And we continue to use the power of simulation during operations to try out new scenarios in the digital world. Conclude they are a benefit before physically implementing them, thus saving time and money. The true benefit of this digital plant: Manufacturing capacity increased by 200%, productivity by 20%.

In partnership with Nvidia, we will now take the Digital Twin, this key technology of the Industrial Metaverse, to the next level. We are creating digital twins that allow interaction in real time, that will look like real machines and, thanks to Siemens know-how, will physically behave like one in every way.

Figure 2: NVIDIA Omniverse and Siemens Xcelerator Interconnected Platforms.

This means that the user does not only see a perfect animation, like in a movie. The twin simulates behaviours just like in the real world. For example, if a rise or fall in temperature is simulated, the digital twin will react to it immediately and in the same way as the real asset on the factory floor or in the building would. Or it can simulate exactly what happens when an individual part fails or new, improved components are installed.

The photo-realistic, physics-related, and real-time digital twin is not the only precondition for the rise of the Industrial Metaverse, but it is possibly the most crucial one. Because digital twins provide the virtual environment for persistent simulation, live interaction and can even serve as a kind of cross-roads between the virtual and the real worlds.

The digital twin is perhaps at the heart of the Industrial Metaverse. But it alone won't suffice to fill the Metaverse with life. We are going to need many additional technologies and innovations. We will need a lot more computing power, faster networks. We will need even more powerful *Artificial Intelligence* (AI), better *Central Processing Units* (CPUs), and more advanced virtual reality technologies.

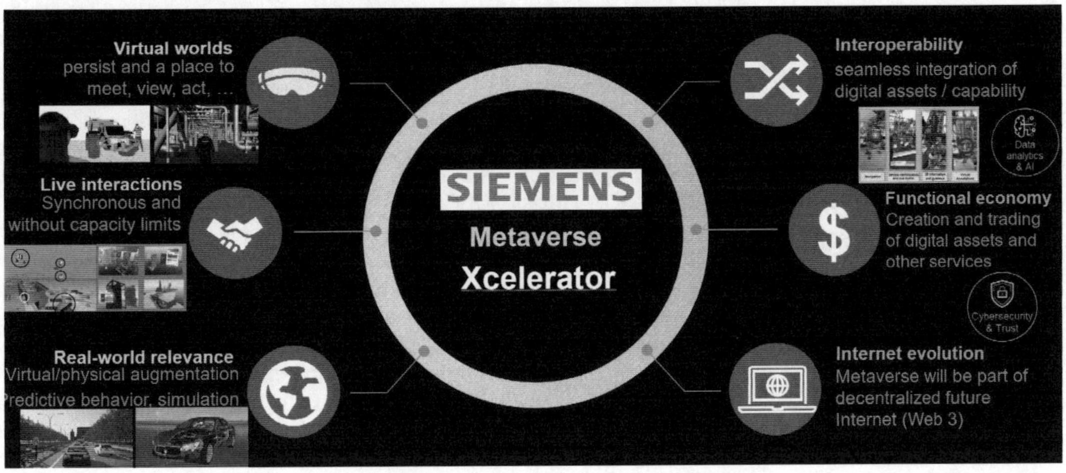

Figure 3: SIEMENS Metaverse in the Cloud Platform.

And these and many other innovations are on their way. We are seeing huge advances that will enable us to build, explore and finally cultivate the Industrial Metaverse. The *Internet of Things* (IoT) already connects more than 10 billion devices and is set to triple in this decade. And with the roll-out of 5G and integrated space-air-ground networks, high-connectivity and low-latency networks will be available around the world. And from virtual and *Augmented Reality* (AR) glasses to edge and cloud technology all the way to blockchain and artificial intelligence, more and more technologies are coming together, that will enable the Industrial Metaverse.

However, it is also clear that this virtual world is not built by just one company or a few partners. It is the users who will create it; it is their reality and experience, their needs and innovations that will shape it. Openness and interoperability are therefore the basic prerequisites for building the Industrial Metaverse. Those who believe they can lock others out will only lock themselves in. And at the same time, people and companies will need easy and flexible solutions to develop and cultivate the Metaverse. Large corporations can perhaps afford to develop tailor-made, individual solutions. Everyone else needs hardware and software solutions that are easy to adapt and preferably work as "*plug and play*". And they need a digital business platform where they can find these solutions quickly and easily. Siemens Xcelerator, for example, will serve this purpose. It simplifies digital transformation and will help pave the way to the Industrial Metaverse. Its solutions are easy to integrate, work together seamlessly and can be adapted very precisely to specific, individual needs. Your digital transformation can start with one machine, one line or an entire factory. And businesses can find a curated portfolio of software and IoT enabled hardware with applications across all industries from many certified providers. You find what you need - and, by the way, you only pay what you need, when you need it. Because all offers come as-a-Service.

By the end of this decade the Metaverse is expected to be a multitrillion-dollar market - with a significant share of industrial applications. But even more important, it can be one of the greatest forces to drive both sustainability and the digital transformation of businesses and entire industries. It will make innovation easier, progress faster, time to market shorter, reduce waste and help us to use fewer natural resources. By adopting the art of the possible with Simulation driven design innovation. You can engineer out as many issues as possible as early as possible but validating every engineering decision. This will help to develop better products by empowering people to explore more alternative designs in shorter time and at considerably lower costs. And it will lead to the easy integration of recycling and circular economy principles into the design process and more efficient ways of production - paving the way for greater efficiency and dematerialization of our economies.

For companies, this next stage of digitalization will, of course, also be a challenge. But at the same time, it is an opportunity to build and cultivate a virtual world that will help solve real-world problems: to become more sustainable trying out alternative eco or recycled materials that are more acceptable to consumers, more efficient, resilient, and more competitive than traditional materials and techniques. This is not only about gaming or social media, but this is also about transforming all industries.

3. METAVERSE AND SUBMARINES

The Metaverse is the blend of the different technologies, some of them has a direct use in Submarines missions and operation, some other would change completely the way we conceive and build them.

Let's review some of the main technologies and how they would affect the design, production, maintenance, operation and lifecycle of Submarines and the industry that backup them.

3.1 CONNECTIVITY & COMMUNICATIONS

It is a key factor and unique in the case of submarines that shall be undetected on deep waters where communications are complex, after all the main defence of a submarine is to stay stealth therefore it may receive information but rarely will expose its position by emitting signals or making communications regularly, the same applies to unmanned underwater vehicles making necessary the ability to have enough "intelligent" and be self-sufficient most of the time. In some cases, is it possible to stay stealth using short distances means like laser but this requires a network of buoys and other vehicles. Another important development is the quantum communications, this will increase the level of encryption without losing speed in case the signals are received by the enemy will be useless. Most of today's communication are still made in low radio frequencies from stations located around the globe but they can send only small amount of data mostly to request that is time to get closer to the surface and contact with the satellite network, it is a risky procedure for the submarine therefore must be used with caution.

The *North Atlantic Treaty Organization* (NATO) Science and Technology Organization's *Centre for Maritime Research and Experimentation* (CMRE) has developed a standard for underwater acoustic communications called JANUS, which is recognised as a NATO standard by all NATO Allies since 24 March 2017. This marks the first time that a digital underwater communication protocol has been acknowledged at international level and opens the way to develop many exciting underwater communication applications [8].

Digital Underwater Networked Communications developed by CMRE provide NATO forces with the capability to gather intelligence, communicate and coordinate underwater and ashore using underwater communications [8].

Figure 4: NATO's Digital Underwater Networked Communications.

The explanation of the number in Figure 4 are:

- **1**. NATO Research Vessel Alliance coordinates and interacts with land-based command and control centre(s) and with underwater vehicles via gateway buoy(s) [8].
- **2**. Gateway buoys communicate with underwater robots via acoustic modems and at the surface via traditional radio waves, through *Wireless Fidelity* (WiFi) or other [8].
- **3**. A submarine can interact and exchange information with friendly forces using the underwater communication network [8].
- **4**. Underwater robots collect and exchange data, communicate with each other thanks to acoustic signals and can work in mission-based teams [8].
- **5**. Unattended sensors may also participate in the network, working together with the robots or simply transmitting data collected over time [8].

3.2 CAD/CAE/CAM AND DIGITAL TWIN

Beside the already mentioned in the point 2. THE "INDUSTRIAL METAVERSE", Submarine may take advantage of the latest developments in digital 3D printing technologies and generate its own supply of spare parts as required. Submarines most probably would become more self-sufficient operational platforms that will handle small fleet of drones and special forces that will require maintenance if they would have to stay longer time on patrol as is being forecast.

Digital twin is essential part of the metaverse, it would predict maintenance of every component, would prevent possible failures attending diverse sensor data, would train personnel and most importantly, would train the AI of the ship itself in simulated operational scenarios as we will comment in the VR section. Submarines are one of the most complex man-made machines and are taking extensively use of this technologies already available.

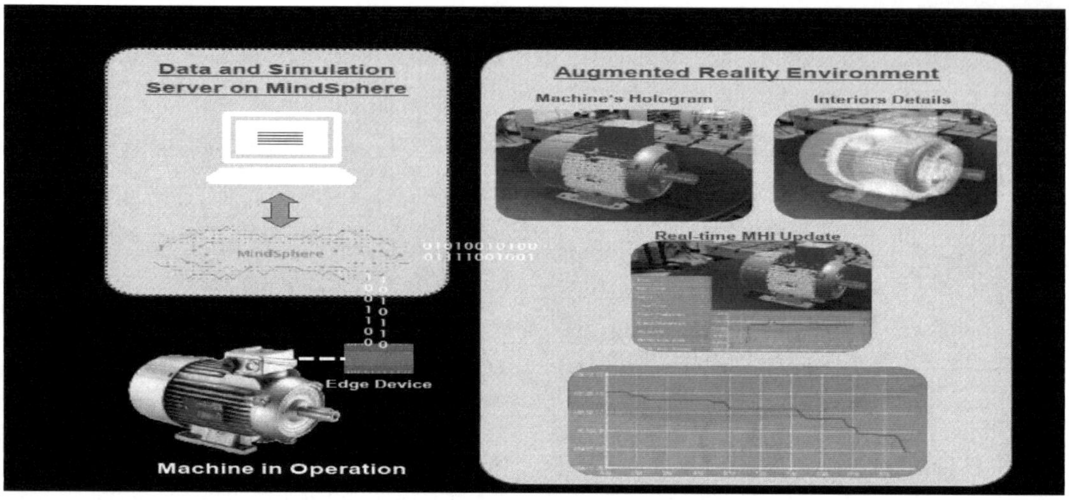

Figure 5: Example of AR and Machine Operation Data Integration.

3.3 AUGMENTED REALITY

AR would facilitate the daily activities of the crew, fast identification of any systems, guided maintenance services, interact and collaborate with intelligent robots, assisting in rutinary and critical missions. In the future the AR will be the ideal tool for humans to interact with the intelligent machines, including the ship main systems.

Taking into account that the machines and systems would be far more intelligent than many humans, the way to leverage the situation would be by enhancing the human capabilities with the AR.

3.4 VIRTUAL REALITY AND SIMULATIONS

VR is essential part of the Metaverse in which we would see the most extraordinary evolution not only because will be key to training the personnel as well as would facilitate the life on board providing a more effective working and relaxing environments, don't forget that a war ship and most particularly a submarine, has reduced spaces available and is submerged in silence most of the time, in this way the crew would be able to walk through the forest, stay in a beach, read a book in the Oxford Library or exercise in a simulated open spaces. As well as would interact with other humans that maybe in another compartment of the ship, in another manned or unmanned vehicle or part of the coordination team of a fleet of submarines and surface ships. In the same way VR would have a more relevant objective and this is to train the ship's AI within different scenarios including war situations and interact virtually with real humans. These simulated exercises would be more relevant than ever due the disruption that the AI would bring to the warfare and the new roles that AI and Humans would play in the future warfare.

VR and Simulations will also be essential part of the training of the Smart Shipyard, where we will see more AI apply to the different shipyard processes (heavy transport, automated store houses, quality control, welding, etc.) and were robots and machines would also be trained to optimize and enhance the quality of may processes.

3.5 INTERNET OF THINGS

The information coming from al the onboard sensors will be handled by the submarine AI itself and would provide key information to engineers and commandants that will improve simulations, predictions, and future submarine development.

The integration of the IoT and the Metaverse will unlock an effortless experience in complex environments with a minimum training. This will support the necessary superhuman powers that we will require to interact with the new generation of intelligent machines to come.

3.6 WEB3

In brief, WEB1 was able to read and present document online, the current WEB2 can read and write information with some level of interaction, WEB3 will be able to enhance all these and, most importantly, will handle and protect the "intellectual property". This enhancement will allow the owners of the information to be identified and apply their rights of intellectual property, securing the information and making the life of hackers more difficult. Today's wars are also been flighted in the battle ground of internet.

3.7 PERSISTENT, SYNCHRONOUS AND ACCESSIBLE

One of the fundamental principles of a Metaverse is that it must be persistent: it cannot be unplugged, restarted, or reset. This is particularly critical in a war scenario where the reality must be reflected in real time as every mission itself, the Metaverse is inherently synchronous and this, even do some events can be programmed and occur autonomously. Finally, as a 3D virtual space, it eliminates all types of barriers, physical or otherwise. It is an endless space where there is no limit to the number of synchronous users, the types of activities taking place, the participating industries, etc. It shall be more accessible than the current Internet platforms available.

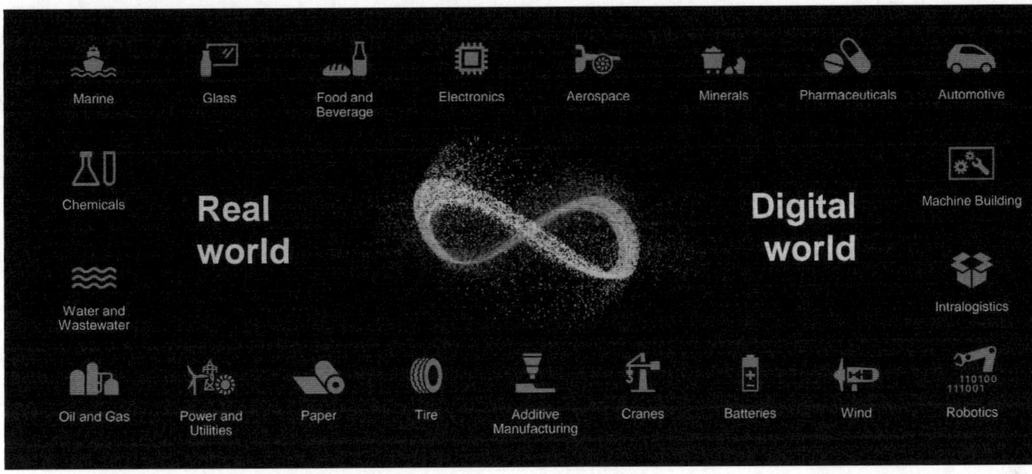

Figure 6: Data integration combined with deep domain know-how for each industry.

3.8 ARTIFICIAL INTELLIGENCE

AI is surely the most important part of the metaverse, it is the technology that will enhance by far our capabilities to understand and improve our world while can be at the same time the most dangerous. The metaverse is the best training ground for the AI.

AI would power operation, design, and smart production in the metaverse prior to the real world and will speed up the future new and more efficient ships.

There are already test that uses AI to support the design on new ships speeding up the design and the evolution to a better and more powerful solutions. To do this it is critical to access the information of previous designs and train correctly the AI, it is easy to understand now the immense role that PLM tools will assume to transfer the know-how to the Machines.

AI also would take care of many services onboard like crew healthcare, pharmaceutical prescriptions, basic treatments. All crew will be constantly monitoring, the ship AI would take care of those that exceed the regulated stress or other possible health problems ahead of developing the external symptoms and avoiding complicating health situation on board (Covid, Surgery, Cancer treatment,…), we must consider that the actual missions would take months underwater, and it is forecast that most probably these times would be increasing in the future.

Considering the huge "drone fleet" that would probably escort the Submarine, the most probable scenario will be a combination of self AI driven drones protecting and supporting the submarine.

In addition, the AI will transform the way we are interacting with the computers in all the fields, there are several example that point our that they would be more like a co-pilot for many activities therefore in a no far future the interaction with the computers and other systems would be more like a "collaboration" partnership, one more team member.

3.9 MERGING OF VIRTUAL AND REAL WORLDS

The metaverse is a world where we would test and train not only humans but robots and their AI this is one of the reasons why we need to simulate as realistically as possible every single detail behaviour with their metaphysics. It will be an environment where we will be commissioning complex software and hardware ahead of is real development. This process will accelerate the development of products, beside the impossibility to recreate some scenarios in the real world, don't forget that the most important mission of a submarine is to deter and avoid those missions for which they are made.

Siemens Xcelerator is an open digital business platform that empowers customers to achieve their desired business outcomes, including growth, increased efficiency, and fast time-to-market through one go-to-place for digital and sustainability transformation - faster, easier and at scale.

It comprises a state-of-the-art, curated digital portfolio; a continuously growing, powerful ecosystem of certified partners; and an evolving marketplace, to explore, educate and exchange with a community of customers, partners, and experts and to purchase products, solutions and professional services like consulting, integration, and financial offerings.

Siemens Xcelerator is a business platform, not a technology platform. It is a place where different parties come together to engage and interact, searching for information, matching demand, and supply, buying from the offerings catalogue, selling their offerings as partners, reaching out to the community for help, meeting to collaborate or co-innovate on a specific project. A platform enabling customers to master their digital transformation easy, fast and at-scale.

Partners can expand their own business by becoming part of a new, innovative platform to share their digital offerings and expertise. Being part of the Siemens Xcelerator ecosystem means strengthened go-to-market and sales activities, increased visibility on an attractive platform, and a future focus on innovation for their own portfolio.

3.10 CYBERSECURITY

The many advantages of an increasingly connected world come at a price: Cyberattacks have become a serious threat. Effective cybersecurity is the order of the day for this very reason. This is also the case for Siemens Xcelerator and its portfolio. Expert knowledge, state-of-the-art technology and the right partners are the important key to creating the best-possible protection.

A hacker attack knocks a local hospital out of action, brings industrial factories and machinery to their knees or sabotages the local public transportation network. These scenarios show: Integrated cybersecurity is the backbone of complex and distributed IoT systems. It is essential for the operation and continued performance of critical infrastructures, production facilities as well as municipalities and communities.

It also plays a key role for companies like Siemens. Why is that? It serves as the operational security of our software-dependent products, solutions, and services. When it comes to things like trains, smart buildings, factories or power grids, there is hardly another company in the world that has set up so many highly sensitive infrastructure projects as Siemens.

- Real-time, photorealistic visualization meets real-world engineering and simulation.
- Immerse users in operations, forecasting performance, for decisions and change.
- Real-time physics predictions.
- Leverage the photorealism of Omniverse to provide physics based decision-making platform.
- Model performance of complex systems in a collaborative environment.

Figure 7: Siemens Xcelerator applied to submarines.

3.11 RENDERING: NVIDIA OMNIVERSE

NVIDIA Omniverse™ is an extensible platform based on *Universal Scene Description* (USD), is helping make it happen by enabling individuals and teams to build custom 3D pipelines and simulate large-scale virtual worlds faster than ever.

The metaverse is the "*next evolution of the internet, the 3D internet, is the internet in 3D, a network of connected, persistent, virtual worlds*" explained by the NVIDIA CEO, Jensen Huang, during an address at SIGGRAPH, the world's largest computer graphics conference.

One example of what can be archive is the following successful business case, Leverage the *eXecutable Digital Twin* (xDT) of a vehicle dynamics model to AV performance with the Omniverse technology.

Our solution uses the xDT technology that Creates a reduced order dynamics model that is self/calibrating and can create virtual sensors. Which reduce their setup time from 3 days to 4h, eliminates expensive sensors, from 40 to 2 gyroscopes while providing the same quality of data. Our result was a xDT that delivers the same level of accuracy.

3.12 PREDICTIVE MAINTENANCE WITH THE XDT

Another successful example was to predict the need for cleaning in industrial *Computer Numerical Control* (CNC) machines. The main challenge facing this customer is that, when the spindle starts vibrating, they cannot confidently define the root cause. In fact, this can be due to chips in the clamping mechanism but also due to suboptimal clamping force on the tool.

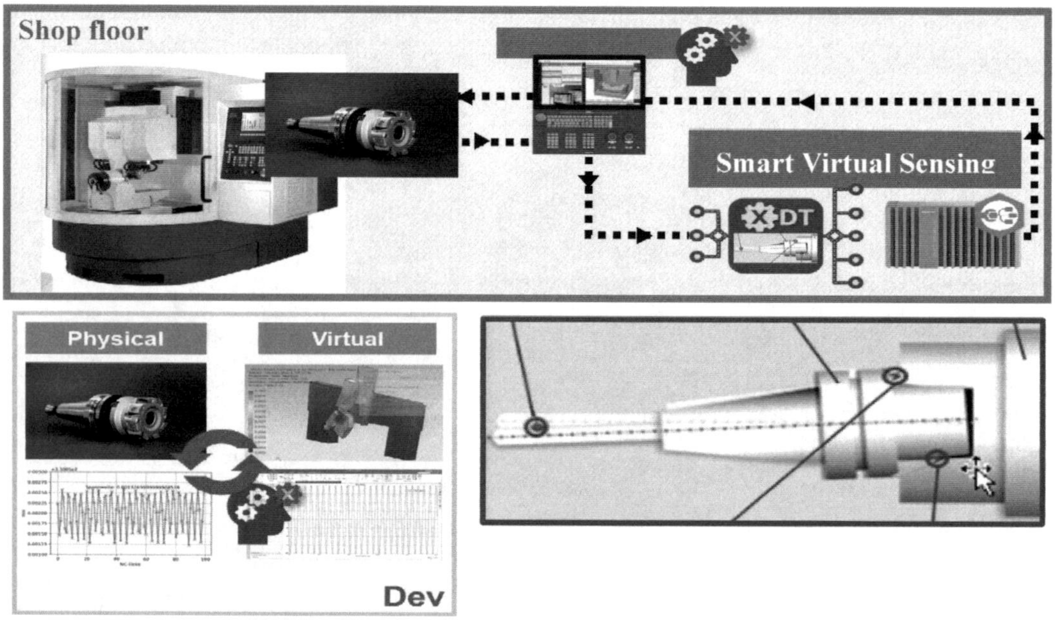

Figure 8: Smart Virtual Sensing xDT will reduce the number of real expensive sensors in impossible places.

Depending on the root cause of the misalignment the company needs a different intervention. With the state-of-use solution is not capable of discriminating between the root causes for the misalignment.

The solution consists of combining simulation and test, engineering out as many issues as possible in the early development phase. To achieve a higher knowledge on the status and performance of the machines. Using Smart Virtual Sensing xDT that is capable of virtually measure the misalignment of the spindle and hence monitor its status during operation constantly giving telemetry feedback. This is then deployed on the industrial edge from DI FA that allows it to run next to the machine and provide direct feedbacks to the control system.

This allows to monitor in real time the misalignment of the spindle while the machine is operating and without the need of expensive sensors that would fail due to the harsh environment.

The main business value here is predictive maintenance where the Smart Virtual Sensing xDT coupled with the existing sensor infrastructure provides additional insights into the manufacturing operation. Rejected parts can be drastically reduced leading to a huge financial saving for the manufacturers.

4. CONCLUSIONS

The Metaverse would transform the future of submarines, they will become more an underwater platform full of unmanned vehicles with a wider scope of missions; recognition, attack, mine hunter, command a small underwater and aerial fleet of drones, special forces, etc. All those mission in collaboration between human and intelligent machines, in this complex world within the Metaverse as optimal operational platform. It will create a more cohesive and collaborative environment.

Submarine crew will work in a VR environment beyond the physical compartments, collaborating with vehicles and robots.

Future submarines would also support this fleet with spare parts mostly fabricated onboard with 3d printers and will stand patrolling instead of moths, by years and behaving like a semipermanent mobile undetected base in the deep of the oceans.

This new marvel of the technology will require previously a fast evolution of the way we design and build submarines that of course shall be integrated in the submarine Metaverse as a first step.

This huge step will require change to the way we see the future and shall be integrate into current platforms and solutions to ensure they fit into the shipbuilding workflows.

Another important aspect of these new world is that optimize maintenance and service activities through virtual asset management, reduce downtime and cost via improved simulation, enhance onboarding and upskill trough immersive and simulated trainings, promote collaborative research & development, improve design and manufacturing by collections real world data from product and production performance and increase awareness.

5. ACKNOWLEDGEMENTS

The authors would like to express sincere gratitude to SIEMENS for providing the support and information for this paper.

6. REFERENCES

1. BALL, M., 'The Meta-Verse; And How It Will Revolutionize Everything', ISBN: 978-1-324-09203-2, 2022.

2. HEILES, J., 'Metaverse Standardization, Overview and Way Forward', *Siemens AG*, 2023.

3. INOUYE, M., 'Evaluation of The Enterprise; Metaverse Opportunity', *ABI research*, Dec.2022.

4. PEREZ, R., and PÉTER, E., 'The use of CAD systems to manage modularity in multi-role warships', *Conference Warship* 2019.

5. ROSSFODDERER, U., 'Presentation: Siemens Nvidia Demonstrator', *Siemens DI SW*, 2022.

6. SIEMENS, 'What is the Industrial Metaverse - and why should I care?', *www.siemens.com*, April 2023.

7. TAMAROZZI, T., 'xDT Fueled Industrial Metaverse, Potential Cases', *Siemens DI SW*, 2021.

8. ROSA, J., 'Understanding the Metaverse: A Business and Ethical Guide', *Publisher: Gildan Media* © 2022

9. North Atlantic Treaty Organization, 'A New Era of Digital Underwater Communications', *NATO.int* © 2017

7. AUTHORS BIOGRAPHY

Rodrigo Perez Fernandez, Ph.D. and FRINA, is Global PFD Director (PES Marine) at *SIEMENS Digital Industry Software.* He is also professor in several subjects at the *Marine Engineering School* of the *Technical University of Madrid.*

Mirko Toman, MBA, M. Sc., currently holds the position of Global Portfolio Development Executive (Dedicated PES Marine Industry in EMEA) at SIEMENS DI SW, his previous experience includes 30 years benchmarking and implementing Shipbuilding CAD/CAM in all sorts of projects and countries.

Warship Conference
31st May – 1st June 2023
Guildhall, Bath, United Kingdom

PRESSURE HULL CORROSION MODELLING AND ASSESSMENT

S Dawe, Babcock SSE, UK

SUMMARY

Corrosion of the steel structure is a major factor in the maintenance of modern submarines. This study tested the feasibility of incorporating laser scan data into a finite element model of a submarine pressure hull. An example laser scan data section was imported and analysed using non-linear finite element analysis. The results were compared to hand calculations, and it was found that there is significant potential to reduce the conservatism in corrosion assessments by utilising this data.

1. INTRODUCTION

The availability of assets is a critical factor across all naval fleets. Many nations have platforms that have been through or are going through periods of commission and life extensions due to delays in long-term maintenance periods, major infrastructure availability and delayed entry into service of new platforms [1]. This presents several challenges for industry as platforms reach the end of the original design life.

Submarines are susceptible to many age-related degradation mechanisms, in particular corrosion of the steel primary structure. Where areas of corrosion are identified as part of regular hull surveys, the extent of the material loss is measured using non-destructive testing methods (NDT) such as ultrasonic thickness measurement. These measurements are then compared against percentage loss criteria set out in the platforms hull survey requirements for assessment by the platform Design Authority. Where the identified corrosion is greater than the acceptable limits, it is typically assessed using hand calculations. These are based on both empirical and first principles analysis and while these methods are swift and widely accepted, they can be conservative when considering localised defects. When considered as part of a major refit period, the decision process and the outcome of a repair decision can be a major contributing factor in the cost of the final repair programme.

One solution to this is to start routinely carrying out more accurate analysis of the individual corrosion defects using finite element analysis to expand the envelope of acceptable defects. Advances in detecting plate corrosion using laser scanning mean we now regularly have access to detailed dimensional data which could be useful if a method of rapidly incorporating it into a model could be developed. This paper looks to test the feasibility of using this data in finite element models of a submarine pressure hull to estimate the resulting collapse depth pressure.

2. FINITE ELEMENT ANALYSIS OF PRESSURE HULLS

The design of submarine pressure hulls presents several challenges to the analyst. As opposed to a tension structure which will stiffen under load, a pressure hull will soften and exhibit highly geometric and material non-linear behaviour. A search of open literature on simulating pressure hull collapse using Finite Element Analysis (FEA) showed that in recent years several researchers have sought to standardise methods, develop partial safety factors [2], [3] and create concept design tools [4]. The collective advice in these papers formed the basis for the analysis method used in this paper.

There have also been several studies that investigate the effect of corrosion on pressure hulls using FEA. A comprehensive study compared results for corroded and intact models to experimental results and found collapse pressures could be predicted accurately as long the eccentricity due to one sided corrosion was captured in the model using shell offsets [5].

3. LASER SCANNING

Over the last 5 years the Non-Destructive Testing (NDT) specialists in Babcock Devonport have been using laser scanning to capture high resolution data for analysis and reverse engineering. For corrosion assessments, point cloud data is converted to surfaces which are then used in combination with ultrasonic thickness measurements to calculate the nominal plate thickness and the percentage loss of material. The scan data provides highly detailed mapping of the corrosion in far greater resolution when compared to using ultrasonic thickness measurements alone. An example image surface is shown in Figure 1.

To test the feasibility of using these images for detailed stress analysis, data from a 1.5 x 0.5m test section of steel plate was used. The plate was scanned using a Creaform HandyScan 700 [6] which has a claimed accuracy up to 0.03mm.

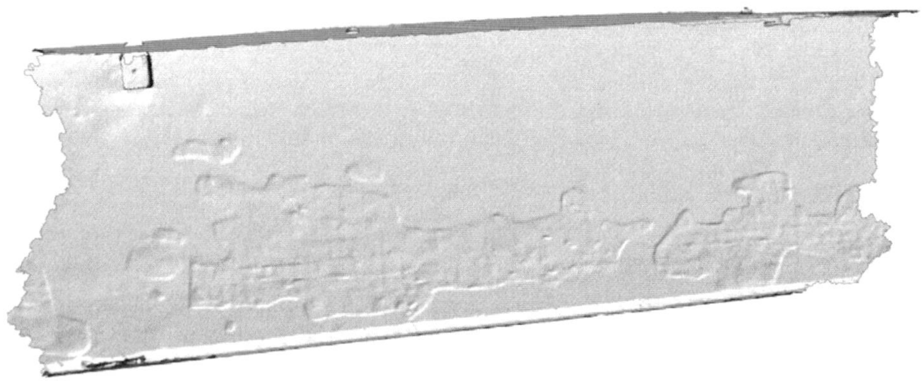

Figure 1: An example of a laser scanned surface as used in this analysis.

4. TEST MODEL SET-UP

For the analysis, a short section of generic pressure hull reinforced with internal T-section stiffeners was created in ANSYS Workbench 2020 R2 [7]. The goal was to create an unclassified geometry which was representative of a typical modern submarine, into which the scanned data could be incorporated (Figure 2).

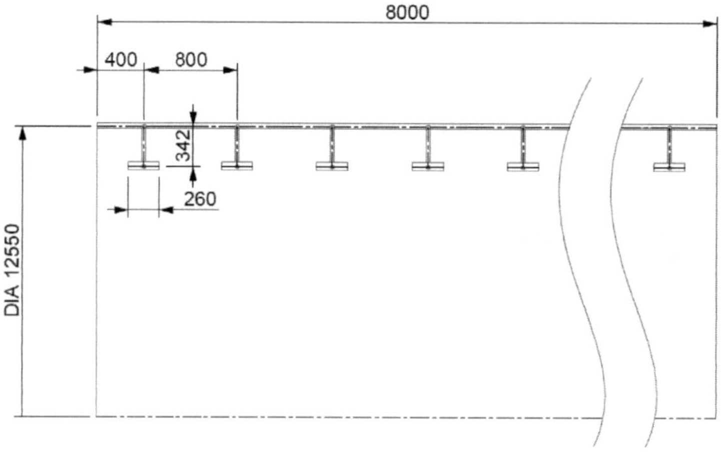

Figure 2: Drawing of the test pressure hull created for this analysis.

4.1 MESH SIZE AND ELEMENTS

The model was meshed uniformly with quad SHELL181, 4 node elements with 6 degrees of freedom at each node. This is a plane stress element suitable for analysing thin to moderately thick shell structures [8].

The ANSYS element reference manual recommends that for a stiffened shell, the most effective choice is to use a reduced integration scheme for the shell and full integration for the stiffeners [8] so this approach was adopted. Where the full integration was used, 5 through thickness integration points were specified to adequately capture the through thickness non-linear effects [2]. An inherent issue with using shells is that unless a method of rigid links or offsets is used at the connecting points, additional material is effectively added at these locations. However, this was investigated by Ref. [5] and the effect was shown to be very small, so this was not accounted for in this study.

4.2 MATERIAL MODEL

The material used for this analysis was Q1(Navy) Steel. This is a quenched and tempered steel commonly used in the manufacture of UK submarines and specified by Def-Stan 02-736 [9]. It is similar to HY80 which is another common steel found in submarines. To improve the accuracy of the analysis data for Youngs Modulus and Poisson's ratio were taken from measured results in Ref. [10].

It is widely recognised that residual stresses can have a significant impact on the predicted collapse pressure of pressure hulls. These are introduced to the structure through both cold rolling of the frames and by welding. Ref. [12] looked at several methods for including residual stress for both Q1(N) and HY80 steel and it was found that using an effective stress-strain curve was accurate and efficient, so this approach was adopted for this analysis.

Ref. [2] suggests a method for incorporating residual stresses from welding by overlaying it on the effective stress-strain curve, however deriving this curve was beyond the scope of this paper. Ref. [11] presents a method for incorporating both residual stress due to welding and cold rolling by using experimentally derived stress values for HY100 steel, however it was shown that the results were largely governed by the geometrical imperfections, and the predicted reduction in collapse pressure due to residual stress was similar in magnitude to the methods presented in earlier papers. Ref. [3] looks at incorporating residual stresses due to welding by modelling the thermal cooling and showed about 1% reduction in the calculated collapse pressure. Given that this was only a comparative study and that the impact was likely to be very small, residual stresses due to welding were omitted.

4.3 HARDENING RULE

The hardening rule determines when the material will yield again if the loading is continued or reversed. Elastic-perfectly-plastic models assume no hardening and therefore the yield surface remains constant. However, given that this model will develop significant plasticity, it was decided that a hardening rule should be implemented.

Two hardening laws are available in ANSYS Workbench: Isotropic and Kinematic. In kinematic hardening, the yield surface remains constant in size but translates in the direction of yielding, as opposed to isotropic hardening where the yield surface expands uniformly in all directions. In anticipation of relatively large strains, a kinematic hardening rule was used for this model as this incorporates the Bauschinger effect. The Bauschinger effect is defined as a decrease in the yield strength of the material in compression because of prior deformation in tension. Ref. [13] found that for a similar study the impact due to Bauschinger was minimal as the materials do not exhibit significant hardening so although accounted for, this was not considered a driving factor in the results.

4.4 BOUNDARY CONDITIONS

Specifying the correct boundary conditions is a key factor in determining the accuracy of an FEA study. For a pressure hull compartment, it will typically be bounded by bulkheads at either end which would be loaded symmetrically either side, resulting in a zero-rotation condition at the bulkhead. For a pressure test scenario of a section, the test would typically include bulky caps bolted to the test piece which again offer rotational restraint at both ends. A review of the boundary conditions recommended in Refs. [13] and [5], found that the consensus was for a 'quasi-clamped' end condition whereby both end rotations and translation degrees of freedom were constrained at both ends except for axial translation which must be permitted at one end. This method was adopted for this analysis by assigning constraints to the edge geometry of the model.

4.5 LOADS

To replicate the hydrostatic loads seen by diving submarines, the model was loaded with both an external surface pressure and an axial end load applied to the edges of the shell elements at one end of the model. The surface loads were defined normal to the element surface while the end load was defined by the global longitudinal axis.

4.6 MESH CONVERGENCE STUDY

To verify the mesh a convergence study was done with the target of 1% convergence on static stress results with a nominal 1MPa load. The starting point was a general 160mm mesh size, followed by 80, 40 and 20mm. The results were benchmarked against the mid-bay maximum circumferential stress which showed that an 80mm mesh size was sufficient to generate accurate results while still solving within a reasonable time (see Figure 3).

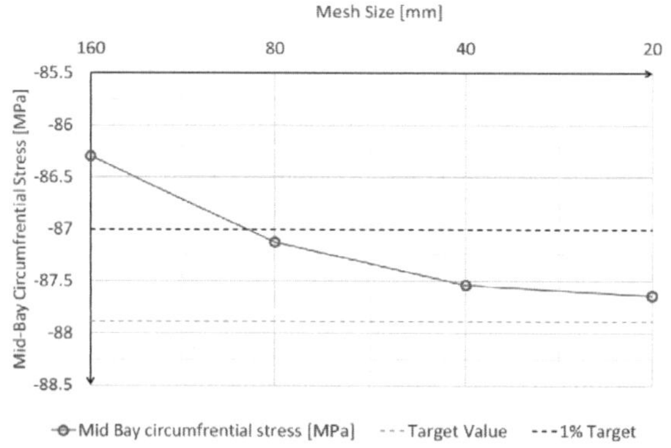

Figure 3: Graph showing the results of the mesh convergence study.

4.7 GEOMETRIC IMPERFECTIONS

The manufactured geometry of any pressure hull will include deviations from circularity and indentation of the plating. For a buckling sensitive structure these geometric imperfections are a key factor in determining the performance of a pressure hull. For analysis, the hull imperfections can either introduced from measured data or can be approximated by seeding imperfections into the model that replicate the critical buckling shapes. A linear buckling analysis showed that the critical buckling mode shape for this geometry was n = 5, so this shape was superimposed onto the model to a maximum deflection of 0.5% on radius as per typical pressure vessel design standards [14] (see Figure 4).

To replicate the typical distortion due to welding (also known as 'hungry horse') the geometry was modified so that between frames the shell was concave to a maximum deviation of 10% of the shell thickness [2] (see Figure 5).

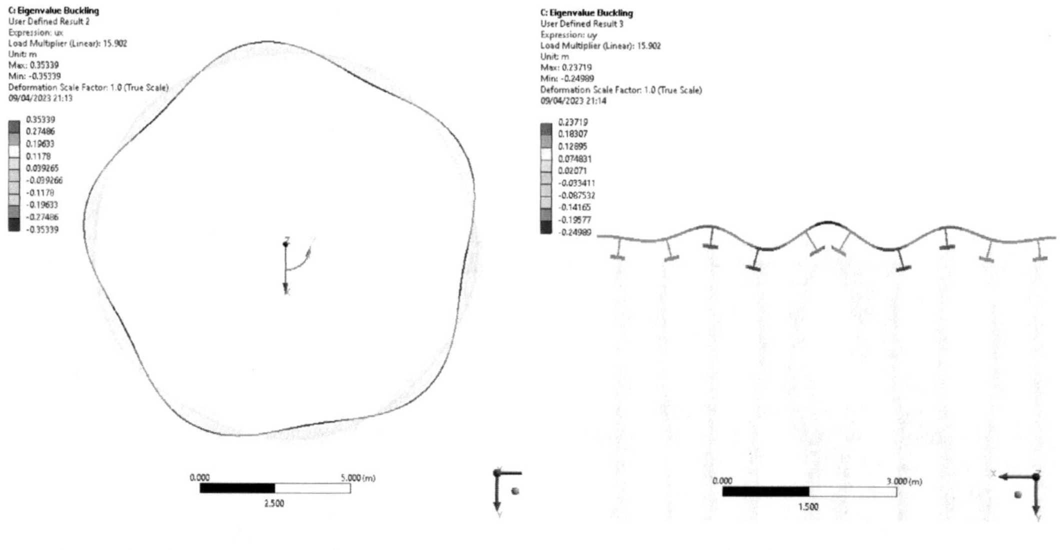

Figure 4: Cross section views showing the lowest calculated critical buckling shape for this geometry.

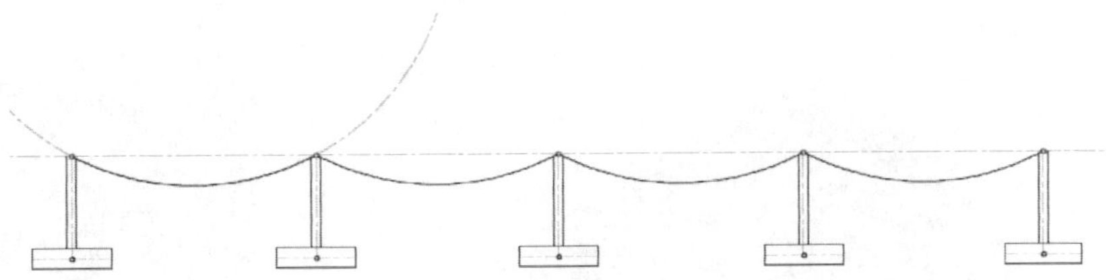

Figure 5: An exaggerated sketch of the imposed 'hungry horse' weld distortions that were imposed on the geometry.

4.8 NON-LINEAR SOLVER PARAMETERS

To predict the collapse pressure for the geometry a non-linear static analysis was run in ANSYS Workbench. The loads were applied incrementally in 20 load steps with automatic time stepping between load steps. ANSYS defaults to a Newton Raphson (NR) method with a 0.5% convergence tolerance (see Ref. [2] for a detailed review of the convergence tolerance) and the 'Line Search' facility was also utilised to aid convergence. The NR method was chosen over the Arc-length method for its speed and ease of implementation. The disadvantage of the Newton Raphson method is that it is not able to predict results beyond the initial buckling of the structure as it fails to converge once the tangent stiffness matrix is singular. However, for this analysis where post buckling performance is not a concern this method was deemed acceptable. Solve times become particularly crucial when multiple scenarios or projections may be required. On a high-performance computer, solve times were typically around 7 minutes. Given that this is a relatively small model, this was deemed an acceptable solve time.

4.9 BENCHMARK DATA

Two versions of the model were benchmarked against hand calculations for overall collapse and the interframe collapse curve in PD5500 [15]. The first version incorporated both the hungry horse distortion and buckled shape described in 4.7 while the other model omitted the hungry horse distortion. The results for predicted collapse pressure are shown in Figure 6, with both models predicted to fail between the interframe and overall collapse predictions made by hand calculation at 4.17 and 4.31MPa respectively.

The plastic strain results from the FEA showed that failure occurred in the mid-bay plating (maximum equivalent plastic strain of 2.25%) aligned with the circumferential buckling lobes which were used as part of the initial deformed geometry (Figure 7).

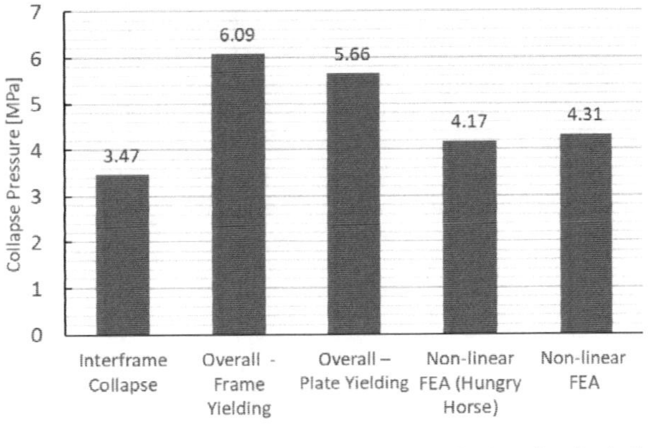

Figure 6: Graph comparing calculated collapse depth pressures for hand calculations vs non-linear FEA.

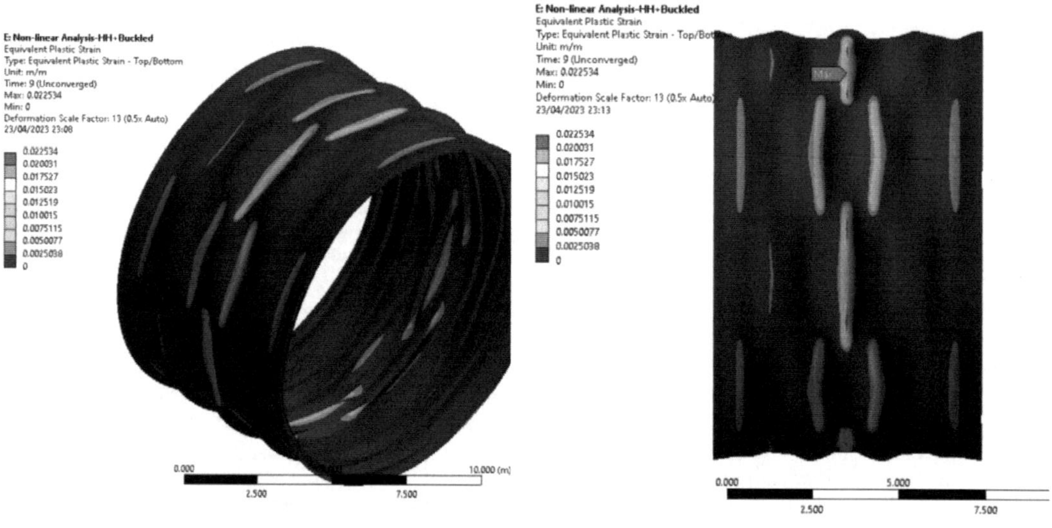

Figure 7: Contour plots of equivalent plastic strain showing failure of the shell plating mid-way between frames.

5. INCORPORATING CORROSION INTO THE MODEL

The advantage of using 3D scan data over manual thickness measurements is the ability to capture the shape of the surface corrosion in far greater detail enabling a more accurate analysis solution. However, the resolution of the data also presents a significant challenge to quickly incorporating it into an FEA model. While experimenting with the example scan data it was found that simply importing and meshing the surface in its original format was impossible. Instead, a lengthy process of simplifying the data was done using ANSYS Spaceclaim. First disconnected, or spurious data points were removed, followed by any non-planar surfaces due to other structure captured in the scan. The corroded regions were then mapped by approximating the boundaries to splines. This resulted in a set of curves which could then be projected onto the pressure hull section for meshing (Figure 8).

A study which looked at predicting the corrosion rate on a HY80 pressure hull showed that if environment specific data is available, it is possible to estimate the impact of not repairing a piece of corroded structure for a given time period [16]. As an extreme case, for this study the regions were assumed to suffer a uniform thickness loss of 20%. While this uniform corrosion approach omits small, pitted regions, it does allow for easily altering the amount of thickness loss. By mapping each region of corrosion separately, it would be possible to assign different levels of material loss to each. It is envisaged that in future studies the thickness profile could be mapped directly from the laser scan data.

The global mesh used in the benchmark analysis required significant refinement in the region of the corrosion to capture the detail adequately without creating poor quality elements. This was done by adjusting the mesh defeaturing, growth and curvature controls within the ANSYS meshing algorithm. This resulted in a very refined mesh in the region of the modelled corrosion (as shown in Figure 9). The corrosion patch was located at the mid-point of the section aligned with the regions of highest plasticity from the benchmark results. Surface offsets were then applied to ensure the internal surface of the shell remained constant to simulate corrosion on the external surface only. For cases of internal corrosion, the surface offsets could easily be altered to maintain the external surface.

Once the mesh was created, the boundary conditions, loads and solver parameters were all set as per the benchmark analysis.

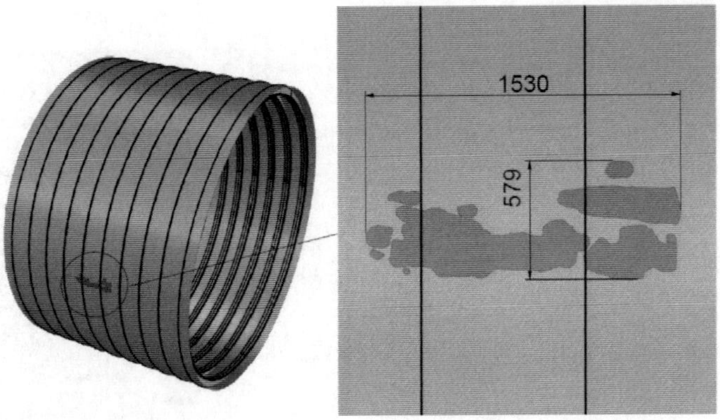

Figure 8: Showing the corrosion site as mapped onto the test model after the simplification process.

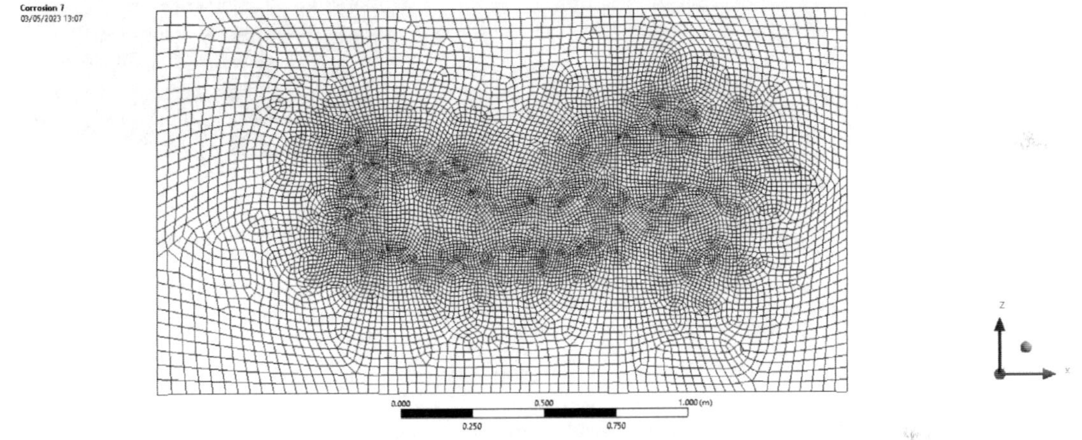

Figure 9: Showing the refined mesh created for the corroded regions.

6. RESULTS

For comparison, the model was assessed using an axisymmetric hand calculation approach to the PD5500 interframe collapse curve, where the corrosion was assumed to act across a whole frame bay. The results showed that the predicted interframe collapse pressure reduces from 3.47 to 2.46MPa (a 29% reduction). Although it is noted that 20% corrosion is an extreme case, it shows that if implemented as a depth restriction on a submarine this would represent a major reduction in capability.

For the linear static FEA case at 1 MPa, the reduced plate thickness area has a major effect on the calculated stresses which increases from 87 MPa to 175 MPa. However, the predicted collapse depth from the non-linear analysis showed a much smaller impact with a final converged load step at 4.08MPa which was a reduction of only 2.16% on the intact result of 4.17 MPa. The mode of failure was similar with plasticity of 7.9% in the region of the modelled corrosion indicating failure of the plate between frames (see Figure 10).

The force-deflection results showed unexpected behaviour with the corroded model showing a steeper force-deflection curve. This slightly stiffer response was attributed to a change in the initial seeded imperfections in the model. By recalculating the linear buckling modes (and subsequent imperfections) taking account of the corrosion, although the general mode shape was the same (n=5), the magnitude of local and global imperfections was slightly altered. Previous studies have investigated the influence of the magnitude of both out of circularity and 'hungry-horse' imperfections and found that the they can have a major impact on the calculated collapse pressure, however it is dependent on the geometry of the pressure vessel [17]. Further investigations into the sensitivity of the model to the imperfections were beyond the scope of this study.

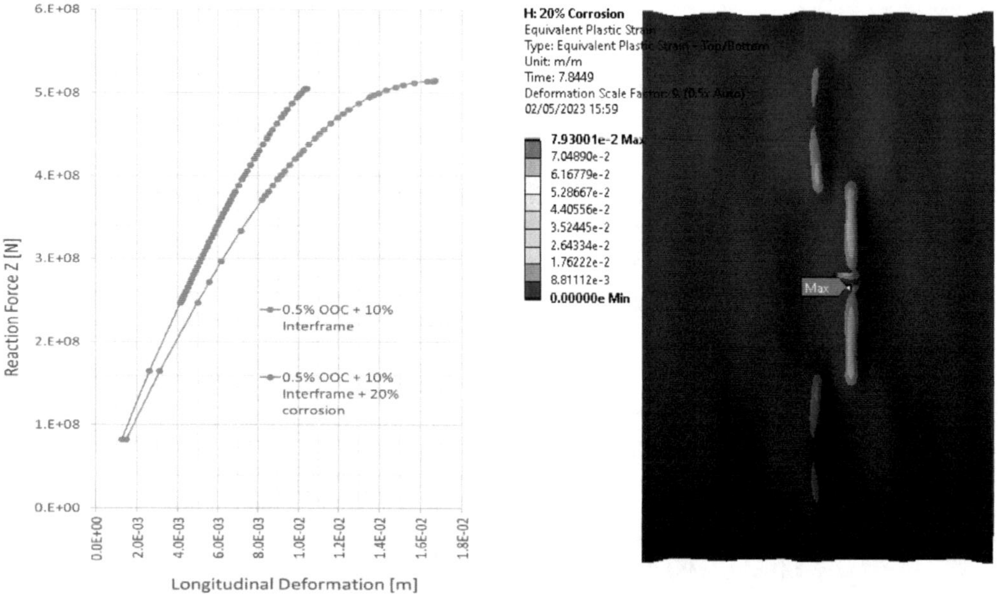

Figure 10: Results of the non-linear FEA showing the force-deflection curves and equivalent plastic strain.

7. CONCLUSIONS

This study sought to test the principle of using laser scanning data for corrosion assessment of submarine pressure hulls using non-linear finite element analysis. A small test section of a laser scan surface was imported into ANSYS Workbench and analysed for stress and predicted collapse pressure.

From this brief study the following points were revealed:

- The manipulation of the laser scan data into a useable format for FEA required a significant amount of work. In order to be useful for rapid assessment, alternative methods should be investigated e.g. a method of mapping the laser scan data onto a fixed orthogonal grid which would be simpler to mesh.
- Once the laser scan data was successfully modelled, it was shown that compared to axisymmetric methods of estimating the impact of corrosion on collapse depth pressure, there is significant potential to reduce conservatism and potentially accept more defects.
- However, it was noted that the results are very sensitive to the input parameters used and therefore standardised processes are needed to ensure accuracy and repeatability.

8. REFERENCES

1. NAVY LOOKOUT, "Getting boats to sea - efforts to improve Royal Navy submarine availability," [Online]. Available: https://www.navylookout.com/getting-boats-to-sea-efforts-to-improve-royal-navy-submarine-availability/. [Accessed 23 04 23].

2. MACKAY, J., and VAN KEULEN, F., "Partial Safety Factor Approach to the design of submarine pressure hulls using nonlinear finite element analysis," *Finite Elements in Analysis and Design,* vol. 65, pp. 1-16, 2013.

3. GRAHAM, D., "A Partial Safety Factor for Pressure Hull Collapse Prediction Using Finite Element Analysis," in *RINA Warship 2008*, Glasgow, 2008.

4. CRAVEN, R., TANNER, D., HUGHES, K., and GRAHAM, D., "A Standardised Approach to Finite Element Analysis in the Design of Submarine Pressure Hulls," in *NAFEMS UK Regional Conference 2022*, Milton Keynes, 2022.

5.

6. MACKAY, J., JIANG, L., and GLAS, A., "Accuracy of nonlinear finite element collapse predictions for submarine pressure hulls with and without artificial corrosion damage," *Marine Structures,* vol. 24, pp. 292-317, 2011.

7.

8. CREAFORM. [Online]. Available: https://www.creaform3d.com/en/handyscan-3d-silver-series-professional-3d/technical-specifications. [Accessed April 2023].

9. ANSYS, "2020 R2 Release Notes," [Online].

10. ANSYS, "Element Reference Manual".

11. DEFENCE EQUIPMENT & SUPPORT, "Defence Standard 02-736 Iss 04 - Requirements for Q1(Navy) Quality Steel," 2020.

12. FARRELL, S., and MACGREGOR, L., "Residual Stress Analysis of Q1N Submarine Pressure Hull Steel with the Portable Miniature X-Ray Diffractometer," *DRDC Atlantic*, 2008.

13. FRANQUETTO, P R., and NETO, M M., "Experimental Residual Stress and Geometric Imperfections on Pressure Hull Instability Analysis," *Journal of Offshore Mechanics and Arctic Engineering,* vol. 140, 2018.

14. GANNON, L., "Prediction of the Effects of Cold Bending on Submarine Pressure Hull Collapse," *DRDC Atlantic*, 2010.

15. GRAHAM, D., "Predicting the collapse of externally pressurised ring-stiffened cylinders using finite element analysis," *Marine Structures,* vol. 20, pp. 202-217, 2007.

16. SUBMARINE DELIVERY AGENCY, MAP 01-074 "Design of Submarine Structures", 2020.

17. BSI, PD5500:2021 Specification for unfired pressure vessels, 2021.

18. VILJOEN, H C., MAHOMED, N., CUPIDO, L H., and MITCHELL, G P., "Effect of corrosion thinning on depth of operation: Case study of an HY80 steel submarine pressure hull," *Marine Structures,* vol. 81, 2022.

19. SHIOMITSU, D., and YANAGIHARA, D., "Estimation of ultimate strength of ring-stiffened cyclindrical shells under external pressure with local shell buckling or torsional buckling of stiffeners," *Thin-Walled Structures,* vol. 161, 2021.

20. MACKAY, J., "Accuracy of nonlinear finite element collapse predictions for submarine pressure hulls with and without artificial corrosion damage," *Marine Structures,* pp. 299-317, 2011.

21. JIANG, L., and WALLACE, J., "Submarine Structure Modelling and Analysis for Life-Cycle Management- Phase 2 Final Report," Defence R&D Canada - Atlantic, 2008.

10. AUTHOR'S BIOGRAPHY

Simon Dawe holds the current position of Structures SME within the Constructive Group of Babcock SSE. He is responsible for ensuring the quality of the technical work from the Structures team and training.

DERIVING SUBMARINE SIGNATURE REQUIREMENTS

WJ Wood, BMT Canada, Canada

SUMMARY

Stealth is critical to the operational effectiveness of submarines, but how does one translate an abstract concept like 'stealth' into a set of performance requirements that can be captured in a System Requirements Document? This paper discusses the wide range of signature types, and some of the challenges that the Signature Manager must face when deriving signature requirements. It discusses previous approaches that have been used, and describes an approach that has been found to be effectively for establishing signature requirements, and which could be used to support decision making throughout the project. This approach would help to ensure that the submarine achieved the optimum performance overall, and could then be used to characterise the operational effectiveness achieved by the delivered in-service platform, in terms that non-experts can understand. However, it ends with a warning that this approach would not be suitable for all signatures in all scenarios, and advocates the acceptance of a pragmatic alternative when necessary.

1. INTRODUCTION

The purpose of this paper is to present some lessons learned from personal experience of deriving signature requirements for submarines. Although it does not provide anything that is fundamentally new, it describes an approach which could be used to produce robust signature requirements that are traceable back to operational needs and explains why this is not always possible and what should be done instead.

The opinions expressed in this paper are based on personal experience and are solely those of the author.

2. THE IMPORTANCE AND CHALLENGES OF SIGNATURES

Stealth is the most important capability that submarines possess. Without stealth, a submarine is a badly designed surface ship, a target and a liability. With stealth, it is arguably the most potent force at sea. Since their inception, a massive amount of effort has been invested in making submarines increasingly stealthy, and this is only matched by the work to develop novel and increasingly capable ways to detect them. Today, detection systems exploit a wide range of signature types, and for a submarine to be truly stealthy it must minimise and control them all.

Some of the better-known submarine signature types are:

- Underwater Radiated Noise (URN)
- Target Echo Strength (TES)
- Magnetic
- Electric
- Radar Cross Section (RCS)
- Infrared (IR)
- Visual
- Pressure

It is clear from this list that those involved in signature control systems must be knowledgeable across a wide range of different physical domains. Many of these signature types have multiple contributing components, each with its own sources and methods of signature control. Effective control of these signatures takes a great deal of ingenuity and attention to detail, and the need to maintain that attention to detail often reaches across every corner of submarine design.

However, at the heart of submarine design there is always the need for compromise. The sizing and limits of the submarine (e.g. displacement, power, cooling etc.) are established very early in the design process, and once this has been done every decision that is made has either a direct or indirect impact on every other part of the submarine. Although signatures will always be critical to success, stealth alone would not be sufficient to deliver an effective submarine, and a balance must be struck between making submarines as undetectable as possible and delivering the most capable submarine overall.

3. WHY SETTING SIGNATURE REQUIREMENTS IS DIFFICULT

Writing requirements correctly is a skill in its own right. While most submarine requirements can be stated in simple, measurable terms that can be traced back to international standards or fundamental physics, this is not true for signature requirements. Submarine signature performance is a highly sensitive area, and nations do not share this type of information willingly. Therefore, there can be no international standards. Signature requirements can vary widely, as they must reflect the operational needs of the navy, which may be different from nation to nation. It should also be remembered that the performance that a submarine needs to be effective will also depend on the capabilities of the threats that it must face.

This can make it extremely difficult to define signature requirements in terms that are simple, robust, meet all of the needs of requirements engineering and that can be used as the basis for Qualification and Acceptance of the submarine.

The approach to submarine design and build has evolved greatly in recent decades, and projects have tried several ways to express signature requirements. Some of those approaches are presented in the following sections, with a brief discussion of their suitability.

3.1 THE SUBMARINE MUST NOT BE DETECTABLE

Although this is obviously what everyone would like, it is also possibly the worst type of signature requirement. Mainly because it is not physically possible, and so no designer should ever accept it. Although great strides have been made over the decades to improve stealth, submarines are still physical objects and there will always be a point at which they can be detected. Even if that distance became very small for some signatures, others would then become the limiting factors. Furthermore, the efforts to make them undetectable might place huge impositions on the design of the submarine, that could make it totally ineffective in any type of practical submarine operation.

3.2 SIGNATURES MUST BE MINIMISED

This type of approach has been used quite successfully in the past, but mainly for those projects where government agencies retained responsibility for the resulting signature performance. It has the virtues of being physically possible and (if appropriately managed), could encourage the design teams to work together to achieve the best possible outcome.

However, it also has some serious drawbacks that would make it totally inappropriate as a contractual requirement to be placed on a supplier. This type of requirement is totally subjective. In the absence of a measurable performance level, there would be no way to verify or validate whether the requirement had been met and the supplier would be free to interpret it any way that suited them. There would also be the issue that each member of the submarine design team would see this requirement from their own unique perspective, and there would be no clear basis for agreeing when the appropriate balance between signatures and other submarine capabilities had been achieved.

3.3 SIGNATURES MUST BE AT LEAST AS GOOD AS THOSE OF THE PREVIOUS CLASS

This approach is a distinct improvement on the previous suggestions and has at least two major advantages from the designer's point of view. As it is based on what has already been delivered, there should be little argument that the requirements are achievable. Also, if the relationship between the design of the previous class and the signatures that it achieved are understood, then the designer should be able to introduce improvements relative to that design with relatively low risk.

The drawbacks of this approach are mainly associated with the need for the submarine to maintain its operational effectiveness throughout its service life. The capabilities of the threats will have evolved since the last class was built, and the new signatures may need to be significantly improved to keep pace with them. This issue is also made worse by the length of submarine design and build programmes, and a decades-long service life. Although the requirements might include an arbitrary margin of improvement to allow for an improved threat, there would be no guarantee that this would be sufficient. Conversely, if that margin was excessive, attempting to meet it could compromise other important submarine capabilities, and the overall operational effectiveness of the submarine.

3.4 SIGNATURES MUST BE AS LOW AS YOU NEED THEM TO BE

Although this approach might sound far too simplistic, it is actually ideal in terms of achieving what is needed without compromising the operational effectiveness of the submarine overall. Of course, the requirement could not be expressed in so general a manner. Instead, it would need to be written in measurable terms, specifying the signature performance that the submarine would need to be effective. The critical question would then become 'how do you establish what you need?'

4. NEEDS-BASED SIGNATURE REQUIREMENTS

Ideally, there are three basic steps to establishing the signature performance required to satisfy operational needs:

- Establish the operational needs of the submarine.
- Calculate the signature levels needed to meet those needs using physics.
- Account for the effects of real-world issues.

4.1 ESTABLISHING THE OPERATIONAL NEEDS OF THE SUBMARINE

The first step in establishing signature requirements that are based on operational needs, is to determine what those needs are. The role that the Navy would need to play in supporting the Defence needs of the nation, should already have been established in high-level government doctrine. Naval planners will then analyse how the various assets at its disposal would be deployed to fulfil that role, and the types of mission that the submarine would need to conduct.

Submarines are extremely versatile platforms, and the particular needs of every nation are different, but examples of some of the roles that are common to many submarines include (but are not limited to):

- Anti-Submarine Warfare (ASW)
- Anti-Surface Warfare (ASuW)
- Intelligence, Surveillance, Reconnaissance (ISR)
- Support to Special Operations Forces

The role of the Signature Manager is to establish the signature levels that would enable the submarine to be successful in the missions that it could be called upon to undertake. This begins by developing a set of highly descriptive mission scenarios that include assumptions about the submarine, and the nature and performance of the threats (e.g., submarines, surface ships, aircraft etc). These scenarios would also need to include assumptions about where the missions would take place, and those assumptions would lead to further assumptions about the environmental conditions that would be encountered.

It would also be important to include realistic descriptions of how the submarine would be operated during the encounters. Those descriptions should be informed by experienced submarine operators, and the value of operator experience should never be underestimated. Submariners understand the value of stealth, are the best possible source of the detail needed to develop the mission scenarios and are best placed to perform an all-important 'sanity check' of the scenarios overall.

Consideration would also need to be given to what might happen during the service life of the submarine. It is generally accepted that modern submarine projects take around fifteen years to complete, after which those submarines would be likely to remain in service for at least 35 years. Therefore, the scenarios would need to consider not only the current geopolitical situation and threats, but also how they might evolve over the next half a century! This requires inputs from assessments made by defence analysts, based on a combination of Defence Intelligence and Defence Science to predict the future challenges that the submarine might face.

Once the scenarios are sufficiently detailed, the operational needs of the submarine can be quantified in terms of the performance of each of the critical systems. The key to successful submarine operations, is the ability to maintain situational awareness of all potential threats in its vicinity whilst enjoying freedom of action and remaining undetected. This means that the submarine must be able to detect, classify and track contacts of interest at a range significantly beyond that at which threats could counter-detect the submarine. This is known as 'range-advantage', and it is fundamental to setting practical signature requirements.

4.2 PERFORMING PHYSICS-BASED CALCULATIONS

The range at which a submarine can detect a threat depends on a combination of the signature of the threat and the sensor performance of the submarine. Similarly, the range at which the submarine could be counter-detected depends on the signature of the submarine and the sensor performance of the threat.

If robust assumptions can be made about the submarine's sensor performance and about the sensor and signature performance of the threat, then the signature level required to maintain the necessary range-advantage can be calculated from physics. The validity of the results of those calculations, depends on how closely the physics-based model reflects the real world, and the accuracy of the assumptions.

Although physics-based calculations provide results in terms of absolute performance, they can only ever be valid for the specific circumstances that are being considered. In reality, there are many factors that need to be taken into account, and these may vary widely. Therefore, it is normal to base the calculations on the most challenging set of conditions possible, on the basis that if the submarine cannot be detected in those circumstances, then it would be reasonable to assume that it would not be detected in others. However, although the use of the 'worst-case scenario' should guard against the requirement being too lax, it brings with it a significant risk of driving excessively stringent signature requirements.

Physics-based calculations alone are not able to account for the dynamic elements of the scenario, which would include factors such as:

- Changes in the way the submarine is operated as the encounter progresses, and how those changes might impact the submarines own signature and sensor performance.
- How the threat might react to an indication of a possible submarine in its vicinity.
- The effect of random chance on the likelihood of success for real-world encounters.

Accounting for these requires an approach that can consider a wide range of variations in the details of the encounter, with the results expressed in a statistical manner.

4.3 ACCOUNTING FOR REAL-WORLD ISSUES

Operational Analysis (OA) (which is sometimes referred to as Operational Research), is the established way to account for the impact of real-world issues on the probability of success for the submarine, and the most effective form of OA utilizes computer simulations and Monte-Carlo modelling. This approach involves conducting large numbers of simulations of the encounter, with minor changes between each simulation, and then analysing the results statistically.

Physics-based calculations remain at the core of OA modelling, and the validity of the results continues to depend on how well those models represent the physics involved and the quality of the assumptions. The great advantage that OA brings is that it can account for the effects of random chance on the encounter, and the dynamic interactions between the submarine and the threats involved.

To enable this type of modelling, the details and assumptions included in the scenarios have to be expanded into a set of highly detailed vignettes that consider (amongst other things):

- The aims of the mission that the submarine is conducting.
- The performance of the submarine systems.
- The tactics that the submarine would employ to achieve the aims of the mission.
- The details of the operational environment.
- The threats that the submarine could encounter.
- The details of the capabilities of those threats.
- The tactics that those threats would employ, and how they would react to the detection of a possible submarine.

A computer simulation of the encounter described by the vignette would then be run. Starting positions would be established for the submarine and each of the threats, and the simulation would proceed in accordance with the pre-programmed tactics according to doctrine. If the simulation determined that the submarine had detected the presence of the threats, it would react appropriately to achieve the aims of its mission. Similarly, if it determined that one of the threats had detected a possible submarine, the simulation would cause it to respond in accordance with its own predetermined tactics.

At the end of the simulation, if the submarine was found to have met its objective it would be considered have been a success run, and if it was decided that it had not it would be classed as a failure. The model would then be reset, randomly selected changes made to the parameters (such as the starting position, course and speed of each of the vessels), and the simulation run again. This would be repeated many times (preferably several hundred to achieve high-quality statistical results), and the number of successes compared to the number of failures to calculate the Probability of Success for a submarine with the signature level that had been assumed for that run of the specific vignette.

Finally, the above process would be repeated as a series of batches, with each batch assuming a different signature level. This would allow the analyst to determine how the probability of success varied with the signature performance of the submarine.

6. THE IDEAL APPROACH

The approach discussed above represents the ideal way to set signature requirements, and can be summarised as follows:

- Establish the roles that the submarine would need to undertake.
- Develop scenarios describing typical missions, with inputs from Defence Intelligence, Defence Science and experienced submarine operators.
- Establish assumptions regarding the environment, and performance parameters for the submarine and the threats being considered.
- Conduct physics-based calculations to determine the signature performance needed to maintain the necessary range advantage against that threat.
- Develop detailed vignettes for each of the scenarios, containing as much detail as possible and including assumptions about the operational tactics of the submarine and the threats.
- Undertake encounter simulations for each of the vignettes, to allow for dynamic interactions between the submarine and the threats.
- Run the simulations a sufficient number of times in Monte-Carlo mode, to ensure a statistically valid result.
- Repeat the simulations over a range of assumed signature levels.
- Establish the probabilities of success and risks of counter-detection for the submarine in each vignette, and how they vary with signature level.
- Establish the signature level that would correspond to an acceptable probability of success for each vignette.

7. LINKING OPERATIONAL EFFECTIVENESS TO PERFORMANCE

One major strength of this approach is that it provides a direct link between the operational needs of the Navy and the performance requirements of the submarine, and this link can operate 'in both directions'. This means that it not only provides a robust means for setting the signature requirements but can also be used to help manage the development of the design throughout the programme.

As already discussed, this approach can translate operational needs that are expressed in terms of probabilities, into requirements that can be expressed in terms of measurable levels of performance. The approach it uses to do this produces results that are quantifiable and repeatable, and maintains a direct causal link throughout. Because of this, it can also be used to assess the impact that any potential or measured shortfall in performance could have on the operational effectiveness of the submarine overall. Therefore, it can also be used to guide technical decision making during the design and build process.

Submarine design and build programmes are huge and costly and once they are underway the consequences of delaying or halting them can be very far-reaching. Therefore, it is most important that any technical issues with the potential to cause major impacts on the programme, are assessed and discussed with appropriate authorities in meaningful quantifiable terms. Although it might be possible to calculate how a particular issue could alter the signature level, it can be a lot more difficult to explain the impact this would have on the effectiveness of the submarine. The approach described above enables shortfalls in signature performance to be translated back into changes in the probability of success for the submarine, and this is something that military personnel can directly relate to.

Put simply, the system would be run in reverse. The modified signature performance levels would become the input parameters for the simulations, and the results would be analysed to determine how much those shortfalls could reduce the probabilities of success for each of the mission types.

Similarly, this approach could also be used to assess the overall operational effectiveness that had been achieved by the submarine at the end of the programme. This would follow exactly the same approach described in the paragraph above, but would also include the measured performance levels of all of the other submarine systems relevant to the simulations.

If the drop in probability of success was found to be small, then it might still be acceptable. Similarly, it was found that a shortfall in a particular signature was compensated for by performance above expectations from a complementary sensor system (or vice-versa), then the submarine might still be acceptable. However, if the analysis showed probabilities of success far below those required for certain mission types, hard questions would need to be asked. The suppliers might be told to find ways to rectify the issues, but if such rectifications were felt to be impractical or failed, the submarine might be prevented from taking part in missions of that type. Whatever the outcome, the ability to 'close the loop' between the operational needs of the Navy, the signature performance achieved and the assessed operational effectiveness of the delivered submarine would be a powerful way to close out a submarine project.

8. THE NEED FOR PRAGMATISM

Whilst the approach described above could be used to address many types of signature requirements, there are some scenarios where that would not be possible. Although the details of the reasons for this vary, they are all associated with situations where the submarine is unable to detect the threat at a range that is far enough to enable the submarine to influence the outcomes of the encounter.

Typically, this might include:

- The submarine might have no practical means of detecting the threat.
- The only practical way to detect the threat might involve active transmissions that would alert the threat to the presence of the submarine, long before it could be detected.
- The submarine might only realistically be able to detect the threat after it was already well within the counter-detection range of the threat.

An example of this might be attempting to establish the visual and infrared signature requirements for the mast heads of a submarine conducting ISR operations in the vicinity of a shore station, or an aircraft operating passive sensors only. Although the submarine would be aware of the shore station, it would have no means of knowing when it was being observed. Similarly, the submarine might have to rely on visual sightings to detect an approaching aircraft, which could be extremely difficult in some conditions. Although active sensors (such as radar) could be used to detect the aircraft, these would provide a beacon for the aircraft to home onto, and so that would not be a practical approach.

In situations like these, the probability of success would largely depend on random chance. Obviously, the probability of success would increase if the signature level was reduced, but how low should it be? Without the ability to influence the outcome, there would be no way to calculate an absolute level that would provide an acceptable probability of success, and it has already been discussed that 'the signature should be as low as possible' is not a viable requirement.

Instead, the recommended alternative would be to word the requirement so that it forced the designer to select the lowest signature option that was still compatible with achieving at least the minimum required performance of the other systems, and thereby maximise the operational effectiveness of the submarine as a whole. To do this, the physical parameters that had the most direct affect the signature level would need to be identified, and reduced to the minimum possible, before the design ceased to be able to accommodate those other key functions that the submarine needed to be effective.

Returning to the example of the visual and infrared signatures of the mast heads, these are both dependent on the material used on the outer surface and on the size and shape of the mast head. Given that it should be possible to mandate an appropriate material for use on any of the possible options, the remaining variable would be the size of the mast head. Therefore, the requirements should specify that the mast heads be as small as possible, while remaining large enough to house the equipment needed to satisfy the critical function and performance requirements of that mast. Clearly this approach relies on establishing a clear understanding of which functions are critical compared to beneficial, but that should already have been determined from other studies.

This approach would also have another significant advantage for visual and infrared signatures. Both signature types are notoriously difficult to assess in practice, as the measured signature levels will be highly dependent on the relationship between the object, the environment and the contrast between the object and its surroundings. Amongst other things, they are both impacted by whether measurements are made during the day or at night, the illumination of the sun (or moon), the sea state, the weather and the background against which they are observed. Therefore, it is extremely difficult to measure them in a way that can be used to prove definitively whether the requirements have been met. The use of a simple quantitative proxy for these signatures (in this case size) makes the assessments trivial and removes the need for highly complex test and evaluation processes.

Although some might feel reluctant to accept this approach when it is first suggested, it is an extremely practical way to achieve the best outcomes for the submarine design overall. There are no good alternatives, and the early adoption of this simple and effective compromise, allows those involved to move on to more pressing matters more quickly.

Finally, it should also be mentioned that there are some signatures where there is no point in setting a signature requirement. That is because they are determined by factors outside of the control of the Signature Manager or the designer, such as the classic example of the pressure signature. There is no question that pressure can be an important mine-activation signature, but there is little that can be done to influence it. The pressure signature is a hydrodynamic effect caused by a combination of the size and shape of the submarine, how fast it is travelling and how close it is to the bottom/sensor. Unfortunately, the size and shape of submarines are constrained by several other critical factors, and it would be impractical to change them in any significant way. Likewise, speed and depth are under the control of the operators, who must optimise them for the mission as a whole. Therefore, pressure is one signature that you must simply accept.

9. CONCLUSIONS

- Stealth is critical to the operational effectiveness of submarines.
- It is extremely important that signature requirements are set that reflect national interests and the operational needs of the Navy.
- Although lower signature levels increase the probability of success, they can come at a cost to other submarine functions.
- Signature requirement must balance the need for stealth with the need to deliver the other important capabilities that the submarine needs to be effective.
- Submarine sensor performance must be considered when determining the signature levels required to meet the operational needs.
- The process for establishing signature requirements should ideally involve high-quality physics-based calculations as part of the Operational Analysis used to determine the performance needed for the submarine to be successful in representative scenarios.
- For those signatures where a reasonable basis for Operational Analysis cannot be established, physics-based analysis informed by operator experience could be used as an acceptable alternative.
- For scenarios where the submarine does not have the ability to influence the outcome, the probability of success is largely driven by random chance, and all that can be done is to ‘keep the signature as low as possible’. In those situations, an acceptable alternative could be to force the selection of the option that minimizes the signature level without compromising other important submarine functions.

10. ACKNOWLEDGEMENTS

The author wishes to acknowledge the support provided by BMT Canada for the development of this paper. He would also like to acknowledge the help, support and education that he has received over the past forty years from the colleagues and friends he has made in Defence industry, academia and the submarine services of the United Kingdom, Australia and Canada. They have educated the author so much, that he now understands that no matter how much he learns there will always be much more to know.

11. AUTHORS BIOGRAPHY

Bill Wood holds the position of Principal Submarine Survivability Specialist at BMT Canada. He is responsible for advising clients on submarine acoustic and non-acoustic signature design philosophies, control measures and through-life management. He is also the Capability Lead for Underwater Acoustics and advises on sonar design and other underwater acoustic issues. He has forty years of experience in underwater acoustics, sonar and signatures. He has lived in the UK, Australia and Canada, where he has worked in academia, public service and industry. His previous experience includes 15 years in the DRA/DERA/QinetiQ organisations in the UK where he was the Business Group Manager and Technical Manager for Underwater Signatures, after which he became the Signatures and Acoustics Manager for the maintenance of the Australian COLLINS Class submarines, and Survivability Assurance Manager for the ATTACK Class (Australian Future Submarine).